THE SCIENCE OF

THE TRUTH BEHIND HERCULE POIROT, MISS MARPLE, AND MORE ICONIC CHARACTERS FROM THE QUEEN OF CRIME

MEG HAFDAHL & KELLY FLORENCE
AUTHORS OF *THE SCIENCE OF STEPHEN KING*

Skyhorse Publishing

Skyhorse Publishing books may be purchased in bulk at special discounts for sales promotion, corporate gifts, fund-raising, or educational purposes. Special editions can also be created to specifications. For details, contact the Special Sales Department, Skyhorse Publishing, 307 West 36th Street, 11th Floor, New York, NY 10018 or info@skyhorsepublishing.com.

Visit our website at www.skyhorsepublishing.com.

10 9 8 7 6 5 4 3 2 1

Library of Congress Cataloging-in-Publication Data is available on file.

Cover design by David Ter-Avanesyan
Cover image by gettyimages

Print ISBN: 978-1-5107-7348-6
Ebook ISBN: 978-1-5107-7349-3

Printed in the United States of America

To Scarlett and Bucky, our furry companions who we'll see again in the clearing at the end of the path.

Contents

Introduction

Some tales are as old as time, like boy meets girl, or perhaps boy climbs the mountain and slays the fearsome dragon to become king. These basic, relatable plots have been formed into fairy tales to teach us how the world works. The murder mystery is no different. Girl murders girl for a chance at fame. Boy stabs father for the inheritance. This genre is built upon a skeleton of tropes, fleshed out with a cast of suspects, scintillating clues, and morbid scenes of death. Agatha Christie was not the first human to devise the detective story but, just as Mary Shelley perfected modern horror with *Frankenstein* (1818), Christie brought the murder mystery out from its literary shadows. The bestselling mystery writer of all time, Agatha Christie turned the idyllic landscape of England, as well as the exotic scenery of the Middle East, into settings rife with betrayal, intrigue, and of course (our favorite part!) cold-blooded murder. The architect of hundreds of fictional deaths, Agatha Christie is a woman with no comparable counterpart. Distinguished, clever, and perfectly British, she is the grandmother to so much of the media we consume today. Without Hercule Poirot, would there be Perry Mason? Columbo? Mulder and Scully? Would shows like *Law and Order* (1990–), *CSI* (2000–2015), and all their spin-offs exist? And don't even get us started on all the mystery writers who owe the Queen of Crime their gratitude.

In this book, we recount our own love for Agatha Christie's enduring legacy. As we uncover the science and history behind her most famous works, we also get acquainted with the woman herself. We hope by book's end, you'll feel like you know her as intimately as you've come to know her Belgian detective.

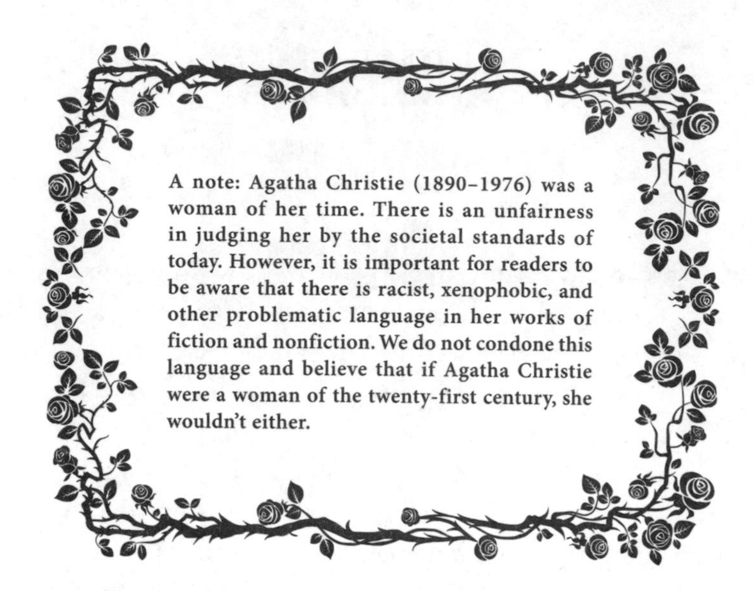

A note: Agatha Christie (1890–1976) was a woman of her time. There is an unfairness in judging her by the societal standards of today. However, it is important for readers to be aware that there is racist, xenophobic, and other problematic language in her works of fiction and nonfiction. We do not condone this language and believe that if Agatha Christie were a woman of the twenty-first century, she wouldn't either.

CHAPTER ONE

The Mysterious Affair at Styles

"One of the luckiest things that can happen to you in life is, I think, to have a happy childhood. I had a very happy childhood. I had a home and a garden that I loved; a wise and patient nanny; as father and mother two people who loved each other dearly and made success of their marriage and of parenthood."[1] So begins Agatha Christie's autobiography, started in 1950 and finished at age seventy-five, fifteen years later. From the beginning of her own story, Christie lived a bucolic yet privileged life that is reflected in the mansion settings of many of her books.

The bestselling novelist of all time, Agatha Christie harnessed abilities that, to this day, make her books a worldwide phenomenon. She artfully unraveled mysteries, creating suspense in the most innocuous of settings. Her well-drawn characters, cheeky humor, and attention to murderous detail has cemented her status as the Mother of Mystery, the Queen of Crime, and the ultimate author of detective fiction.

Agatha Christie in 1964.

We would argue that the reader's ability to follow the clues is what makes Christie's work so addictive. After a novel or two, you become hyperaware of every sentence. A mere mention of everything from a teacup to a muddy footprint gets your mind whirring. Can you solve the mystery before Hercule Poirot or Miss Marple? Or will you be just as dumbfounded as the characters in the book when the true murderer is revealed? On only the rarest of occasions can we keep up with Christie's skill of logic. Her detectives are much too clever for us to outdo.

One subject that Christie touches early on in her autobiography is that of a contented marriage: "I am interested in my parents, not only because they were my parents, but because they achieved that very rare production, a happy marriage. Up to date, I have only seen four completely successful marriages. Is there a formula for success? I can hardly think so."[2] This meditation on happy versus unhappy marriages makes sense in the larger context of both Christie's life and literature. At the time of writing her first novel, *The Mysterious Affair at Styles*, Christie, née Miller, was newly married to royal artillery officer Archie Christie. By the time her autobiography was published, she was with her second husband, Max Mallowan. Marriage (and all its complex trappings) is an overarching theme in Christie's work, including this first novel. The character of wealthy Emily Inglethorp has married a much younger man, much to the chagrin of her family and friends, who believe her new husband is simply a "fortune hunter." There is also Emily's son, John, who is suspected of having an affair by his wife, Mary. These are the conditions surrounding Emily's mysterious death by strychnine poisoning.

Recounted in first person by a friend of a genius detective, *The Mysterious Affair at Styles* is strikingly similar to Sir Arthur Conan Doyle's Sherlock Holmes series (1887–1927), a reference cheekily made early in the novel: "Well, I've always had a hankering to be a detective!" Mary Cavendish reveals to Arthur Hastings. He responds, "The real thing—Scotland Yard? Or Sherlock Holmes?"[3] Like Doyle's Dr. Watson, Christie's Arthur Hastings, a British soldier recovering from his time on the battlefront in World War I, is our connection to this insular, wealthy world of rural England. Hastings is at once an interloper, as he knows the other characters in a minimal sense, as well our reliable narrator. Just as practical Dr. Watson gives us a view into the singular mind of Sherlock Holmes, Hastings is our "normal" mirror that reflects the eccentric detective Hercule Poirot. We trust Hastings's descriptions of characters and events because he is an outsider with no stakes for emotional complications. Of course, this is not always true, as by the end of *The Mysterious Affair at Styles* young Cynthia Murdoch catches his eye, and Hastings makes a rather abrupt marriage proposal.

He is denied!

While Hastings is our bridge from reader to the fictional world, Belgian detective Hercule Poirot is the true star of not only Christie's legacy, but the mystery genre at large. He makes his first appearance in *The Mysterious Affair at Styles*, in what would become his usual manner: a coincidence. Staying in a nearby house with fellow Belgians, Poirot is simply enjoying his semiretirement in the English countryside when he bumps into his acquaintance, Hastings, as well as his old friends, the Cavendish-Inglethorp family. His proximity to murder at the Styles manor is merely happenstance. Though the murderers will be regretful that they didn't have patience for the detective to leave before they enacted the fateful deed! This trope has prevailed throughout literary suspense, as well as in the mystery TV shows of our youth, when sleuths like Jessica Fletcher (Angela Lansbury) of *Murder, She Wrote* (1984–1996) always seemed to vacation where murder was afoot.

Modern readers would be hard-pressed *not* to know some book, TV, or film iteration of Hercule Poirot. In 1920, he was borne into *The Mysterious Affair at Styles* with his distinct vigor, a man of deductive genius like Sherlock Holmes, along with an almost manic energy that would surely disturb Holmes from his serious cogitation. Perhaps disparate to Poirot's humor is his precision of cleanliness, described by Hastings:

> Poirot was an extraordinary looking little man. He was hardly more than five feet, four inches, but carried himself with great dignity. His head was exactly the shape of an egg, and he always perched it a little on one side. His moustache was very stiff and military. The neatness of his attire was almost incredible, I believe a speck of dust would have caused him more pain than a bullet wound. Yet this quaint dandyfied little man who, I was sorry to see, now limped badly, had been in his time one of the most celebrated members of the Belgian police. As a detective, his flair had been extraordinary, and he achieved triumphs by unravelling some of the most baffling cases of the day.[4]

Poirot's fastidious nature has been a trademark of his character not only in Christie's books, but in TV and film portrayals. *The Mysterious Affair at Styles* was adapted for the screen in 1990 for the British ITV show *Agatha Christie's Poirot* (1989–2013). The nearly two-hour long episode mirrored the novel quite closely but added a few extra moments of Hercule Poirot's (David Suchet) particular cleanliness. When stopping to investigate a coffee stain on the carpet, Poirot sets down his handkerchief on which to kneel. There is also an added scene in which Arthur Hastings (Hugh Fraser) experiences a war-induced nightmare.

The *Poirot* TV series is a large part of why Hercule Poirot has become such a recognizable symbol of the mystery genre. Throughout seventy episodes, every Poirot novel and short story of Christie's was faithfully adapted, with the mustachioed detective at the center of the action. In an interview with *Strand Magazine*, actor David Suchet revealed how he prepared to embody such an iconic character:

> I plowed through most of Agatha Christie's novels about Hercule Poirot and wrote down characteristics until I had a file full of documentation of the character. And then it was my business not only to know what he was like, but to gradually become him. I had to become him before we started shooting. I worked very hard on finding the right voice. I was desperate that he should sound French, although he is Belgian, because everybody believes that he is French. I wanted to move my voice from my own—which is rather bell-like and mellow and totally unlike Poirot. I wanted to raise that voice up into his head because that's where he works from. Everything comes from there. My voice is very much in my chest and in my emotional area, but his is up in his head. He's a brain, so that voice had to be raised up and perfected. And then I had to learn how to think like him and how to see the world through his eyes. I had to make his mannerisms and eccentricities not as though they had been put on to be laughed at, but as if they had come absolutely from within that person. I had to make it look real for the audience, yet in a way so that they could find themselves smiling at this strange little man.[5]

In every Christie story the cast of characters are introduced and one by one we begin to form our first impressions of them. We think things like, "could *they* be the murderer? That woman seems suspicious! That man could never harm a fly!" Sometimes we're right but often the author has brilliantly misguided us. How are first impressions formed in the brain? When we first see someone, we judge them on their nonverbals whether we're aware of it or not. Nonverbals include the clothes they're wearing, their height, their weight, and everything down to their hair color. How a person stands and speaks also plays a role in how we perceive them.

First impressions have been known to last for months and tend to be linked to our prior knowledge of personality traits. For example, if a person shows up to a formal dinner party in pajamas, we may perceive that they are lazy. If a person arrives at a job interview late, we may assume that they are not taking the opportunity seriously. First impressions can be wrong, of course, so it's important to follow up with clarifying questions or get to know the person through conversation and multiple exchanges. Or, in the case of reading about characters in a novel, follow through the story to see how your impressions may change.

As the story unfolds in *The Mysterious Affair at Styles*, we start to become suspicious of many of the characters. Can people look suspicious? In a way. Our brains form impressions of people within milliseconds of meeting them based on our beliefs about personality. According to a 2018 NYU study, these impressions have real consequences, including:

> A range of real-world outcomes, from political elections, to hiring decisions, criminal sentencing, or dating. Initial impressions of faces can bias how we interact and make critical decisions about people, and so understanding the mechanisms behind these impressions is important for developing techniques to reduce biases based on facial features that typically operate outside of awareness.[6]

Flashbacks of war affect the war veteran, Hastings. How was this treated in history versus now? Even though post-traumatic stress disorder (PTSD) didn't appear in the *Diagnostic and Statistical Manual of Mental Disorders* (DSM) until 1980, war has affected soldiers for centuries. Even the Bible mentions the trials of battle in Deuteronomy: "When thou goest out to battle against thine enemies, and seest horses, and chariots, and a people more than thou . . . the officers shall say, What man is there that is fearful and fainthearted? Let him go and return unto his house, lest his brethren's heart faint as well as his heart."

It wasn't until the French Revolutionary War and the Napoleonic wars that doctors began to observe how soldiers were reacting to the sound of cannonballs being fired.[7] As battles became more technologically advanced in the coming decades and the field of psychiatry grew, people began to understand the concepts of shell shock, hysteria, and combat fatigue surrounding those on the front lines. Treatment ranged from general anesthetic to electric shock therapy. Currently, PTSD is treated with various medications and several types of therapy including cognitive therapy, which recognizes ways of thinking; exposure therapy, which helps patients face situations and memories they find frightening; and eye movement desensitization and reprocessing (EMDR), which combines exposure therapy with guided eye movements. This therapy can help patients process traumatic memories and change how they react to them.[8]

Cynthia is not waking up in *The Mysterious Affair at Styles*, and the others assume she has taken a sleeping powder. What were the sleep tonics of the past? Since the dawn of time, people have been using substances to help them fall asleep, including alcohol and opiates, while barbiturates became popular at the beginning of the twentieth century. By the 1970s, Librium and other benzodiazepines "were the most widely prescribed drugs in the world."[9] Natural sleep aids such as melatonin, lavender, and magnesium can also be effective.

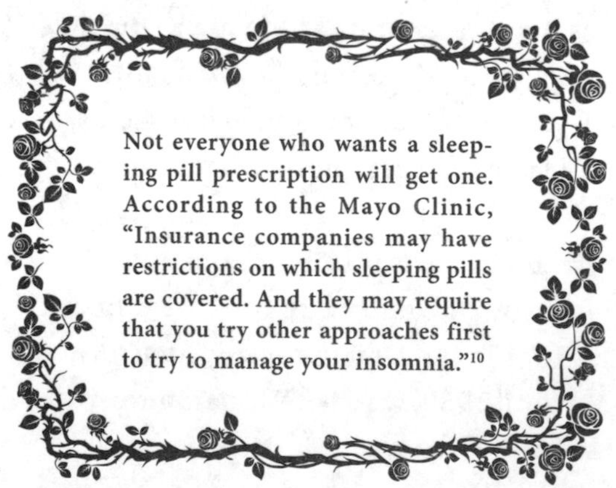

Not everyone who wants a sleeping pill prescription will get one. According to the Mayo Clinic, "Insurance companies may have restrictions on which sleeping pills are covered. And they may require that you try other approaches first to try to manage your insomnia."[10]

Strychnine poisoning occurs in *The Mysterious Affair at Styles*. How does this particular poison affect the body? According to eyewitnesses, strychnine produces some of the most painful symptoms of any toxic reactions and involves muscle spasms, convulsions, and asphyxiation.[11] There is no treatment for strychnine poisoning, and death usually occurs within two to three hours of exposure. Some famous cases of strychnine poisoning include Alexander the Great in 323 BCE[12], who may have drunk wine contaminated with the substance, and Belle Gunness, a serial killer in Indiana in the early 1900s who may have used the poison to murder multiple men.[13]

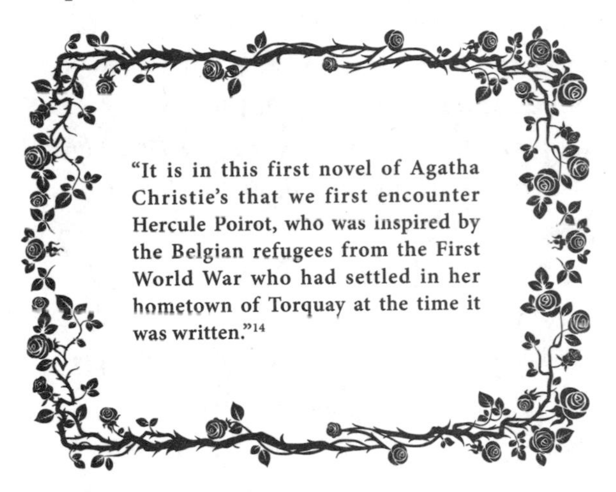

"It is in this first novel of Agatha Christie's that we first encounter Hercule Poirot, who was inspired by the Belgian refugees from the First World War who had settled in her hometown of Torquay at the time it was written."[14]

Poirot talks about the amount of food the victim ingested in *The Mysterious Affair at Styles* and how this little amount could affect the speed at which the poison absorbed into the body. Can food consumption affect the timing and ability of drugs in our systems? Absolutely! This is why some medications are recommended to be taken on an empty stomach while others are meant to be taken with food. According to a 2019 study, "different foods, based on factors such as nutritional composition (high-protein, carbohydrate-rich, or high-fat meals), calorie content (low versus high calorie meals), volume, temperature, and fluid ingestion, have distinct influences on the transit time, luminal dissolution, permeability, and bioavailability of the drug product."[15]

While listening to a witness on the stand, someone says it sounds as if the witness has something to hide. How does the tone of our voice affect our believability? A 2021 study found that humans have a natural prosody to our voices, and we can detect whether a person is trustworthy or not based on listening to them speak. Prosody is defined as the patterns of stress and intonation of language. The study found that people who spoke with less intensity at the beginning of a word and spoke slowly with a variable pitch were thought to be less trustworthy. Those who spoke with a faster rate and placed intensity in the middle of words were found to be more honest.[16] This could be because we think it takes less effort to tell the truth and it takes more time for liars to get their story straight!

Bromide was added to Emily's evening medication in *The Mysterious Affair at Styles*, which ended up killing her. How does bromide affect the body? The long-term consumption of bromide can cause restlessness, confusion, hallucinations, psychosis, and even coma.[17] Lithium bromide was previously used as a sedative in the early 1900s and is still used in veterinary medicine as an anticonvulsant.

The Mysterious Affair at Styles is just the beginning of Agatha Christie's impressive catalogue of works. Chock-full of history, science, murder, and mayhem, there's so much more to explore!

CHAPTER TWO

The Murder of Roger Ackroyd

Like many writers, Agatha Christie did not find immediate success. It took multiple rejections before *The Mysterious Affair at Styles* was published. She was thirty years old. Once her fiction was accepted by both publishers and readers, Christie became an unstoppable force, writing more than sixty detective novels, plays, and short stories over the subsequent decades.

"Author Kate DiCamillo racked up 473 rejection letters in just six years before striking a publishing deal for her first novel, *Because of Winn-Dixie* (2000). Reportedly, literary agents reject 96 percent of writers."[1]

After Hercule Poirot stole the scene in her first novel, Christie's follow-up was a story of espionage told through characters Tommy Beresford and Tuppence Cowley. The couple first starred in her second novel, *The Secret Adversary* (1922), in which they solve a mystery and fall in love. As a married pair with a taste for adventure, they would headline five more works of Christie's fiction. Tommy and Tuppence's hijinks don't hold the lasting appeal of Hercule Poirot or even Miss Marple. Even so, their stories have been adapted on the screen, including *The Secret Adversary* (1929), a silent film.

We have come to focus on *The Murder of Roger Ackroyd* (1926) not only because of its significance in both Christie's life and career, but also because of its brilliance that is still being discovered nearly one hundred years later. As 1926 brought Agatha Christie continued success in her career, her personal life faltered. Her mother, Clarissa Miller, died in April of that year. Christie was left to deal with not only the profound loss, but also the many financial and logistical dealings of her mother's death. In the midst of her grief, Christie was greeted with another nasty life event. Her husband, father to her only child, Rosalind, had fallen in love with a woman named Nancy Neele. They were carrying on an affair right under Agatha Christie's nose. The culmination of this trauma would eventually lead to divorce in 1928.

Despite the ups and downs, Christie continued to write. It became clear that Hercule Poirot was a character readers were curious to follow. He featured in her third novel, *The Murder on the Links* (1923), as well as a short story collection, *Poirot Investigates* (1924).

Narrated by country doctor John Sheppard, *The Murder of Roger Ackroyd* centers on just that: the stabbing death of rich Mr. Ackroyd, a man who had just come into knowledge of a local blackmail plot. And just like in *The Mysterious Affair at Styles*, Hercule Poirot is close by, attending to his quiet, retired life in rural King's Abbot. In fact, he is finding joy in growing vegetable marrows right next door to Dr. Sheppard. The doctor and his sister Caroline (known to be a town gossip) discuss their new neighbor: "My dear Caroline," I said. "There's no doubt at all about what the man's profession has been. He's a retired hairdresser. Look at that mustache of his."[2] Caroline does not accept her brother's appraisal of Poirot. She goes on to tell Dr. Sheppard that she asked Poirot if he was a "Frenchman." Of course, readers of Christie know that the Belgian detective does not like when people assume he is French. This often leads Hercule Poirot to a scowl.

At the outset, this novel has the familiar trappings of Christie's trademarks: Poirot, a close-knit group of suspects, a small town in the English countryside, and the death of a wealthy businessman. Knowing this, Christie twists these tropes to her own advantage, serving what is considered one of her greatest endings to shocked readers. In 2013, the Crime Writers' Association named *The Murder of Roger Ackroyd* the greatest crime novel *ever* written.

I (Meg) have long attested that *The Murder of Roger Ackroyd* is my favorite Christie novel since I read it as a teenager. It delivers such a wallop of an ending that I, much like many Christie fans, was left reeling. Yet, when people new to Agatha Christie ask me what book to start with, I don't recommend *The Murder of Roger Ackroyd* because its impact is felt more viscerally if the reader is well versed in her fiction. It is her usurpation of expectations that makes this book so vital to crime fiction.

Now come the spoilers. (And we really do beseech you, if you haven't yet read *The Murder of Roger Ackroyd*, leave this book, sit down, and read Christie's novel in one pulse-pounding sitting. Then, don't forget to come back to us!) As the novel unfolds, there are many suspects at play. Hercule Poirot investigates them in his typical fastidious fashion, eyeing every detail. Because Arthur Hastings is currently in Portugal, Dr. Sheppard becomes our window to the events. He follows Poirot, puzzling out clues alongside the detective. We, as the reader, are able to eavesdrop on the doctor's conversations with his sister, who holds a lot of town information. And Dr. Sheppard, much like Hastings in *The Mysterious Affair at Styles*, is one of the last people to see the victim, Roger Ackroyd, alive. Poirot, too, sees the similarities of Sheppard and Hastings: "You must have indeed been sent from the good God to replace my friend Hastings," he said with a twinkle. "I observe that you do not quit my side."[3]

This is not the only time Poirot makes reference to Sheppard being his new, or at least temporary, Hastings. The doctor accompanies Poirot on most of the investigation, acting as a sort of sounding board for the detective. This, along with the fact that *The Murder of Roger Ackroyd* is told through the eyes of John Sheppard, makes for a shocking finale. Because the doctor himself is the murderer!

Teasing the very tropes that she herself perfected, Christie shocks the reader by artfully making us question the fictional world she has created. In fact, reading the novel for a second time in research for this book, it was thrilling to recognize the subtle clues that Christie left for the reader. This includes the moment Dr. Sheppard leaves the victim, we assume alive, but are gravely mistaken: "The letter had been brought in at twenty minutes to nine. It was just on ten minutes to nine when I left him, the letter still unread. I hesitated with my hand on the door handle, looking back and wondering if there was anything left undone. I could think of nothing. With a shake of the head I passed out and closed the door behind me."[4] In retrospect this account may sound suspicious, but the Queen of Crime masterfully chooses her words to conceal murder.

The notion of an unreliable narrator existed before *The Murder of Roger Ackroyd*, yet arguably not done in the confines of a traditional crime novel. So, what exactly is an unreliable narrator? "Authors sometimes use an unreliable narrator to tell the story, a protagonist who can't be trusted to tell the events accurately. Either they are insane, evil, delusional, forgetful, or just plain wrong. This is not merely characters sharing different 'points of view.' These narrators purposefully lack credibility."[5] Edgar Allan Poe wrote many prime examples of the traditional unreliable narrator. For instance, in Poe's short story, "The Tell-Tale Heart" (1843), a murderer with a guilty conscience slips from reality to horrific fantasy in telling his tale. In *The Murder of Roger Ackroyd*, Dr. Sheppard is not presented in the same dim light as the narrators in Poe's stories, leaving his unreliable nature to be revealed much later.

The peculiar narrative structure of *The Murder of Roger Ackroyd* made us curious about how it might be adapted on screen. There's a season seven episode of *Poirot* that follows the novel quite closely, yet loses some of its shock factor because of the nature of the visual medium not being told solely through Dr. Sheppard's point of view. Our favorite adaptation is a radio hour recorded in 1939 for the *Campbell Playhouse* (1938–1940), a series in which famous novels were dramatized in hour-length episodes. These ranged from classics like Victor Hugo's *Les Miserables* (1862) to contemporary pieces like Daphne du Maurier's *Rebecca* (1938). Orson Welles hosted *Campbell Playhouse* (sponsored by the soup company!) shortly after his infamous turn on *The Mercury Theatre on the Air* (1938) in which he frightened listeners with a too-real telling of H. G. Wells's 1898 sci-fi novel *The War of the Worlds*.

Orson Welles portrayed Poirot and Dr. Sheppard in 1939.

In the *Campbell Playhouse* iteration of *The Murder of Roger Ackroyd*, Orson Welles plays both Hercule Poirot and Dr. Sheppard so seamlessly that we didn't notice until this was announced at the end! The celebrity guest spot of Caroline Sheppard was given to Edna May Oliver, a character actress best known for her work on Broadway as well as films like David O. Selznick's *David Copperfield* (1935). After Welles's turn as the Belgian detective, he went on to an illustrious career in film, directing, writing, and starring in what is considered by many to be the best film ever made, *Citizen Kane* (1941).

Investigating whether a death was murder or suicide comes up in *The Murder of Roger Ackroyd*. How do investigators tell the difference? Several factors are taken into consideration when first looking at a potential crime scene, including the placement of the body, the placement of a weapon if one is present, the forensics of the scene, and the general layout of the surroundings. For example, if someone was in the middle of a task, like cooking dinner or folding laundry, it may become obvious that a suicide was staged. The arrangement of the furniture in a room becomes a clue later in the investigation of *The Murder of Roger Ackroyd*. Ultimately, the cause of death will help determine whether foul play was involved and needs to be investigated further to rule out homicide.

Other methods that investigators use to determine if someone died by murder or suicide is interviewing family and friends to determine the victim's mental state, finding a suicide note, or discovering recent changes in the person's life. According to Vernon J. Geberth, a homicide and forensic consultant:

Victimology is the collection and assessment of all significant information as it relates to the victim and his or her lifestyle. Personality, employment, education, friends, habits, hobbies, marital status, relationships, dating history, sexuality, reputation, criminal record, history of alcohol or drugs, physical condition, and neighborhood of residence are all pieces of the mosaic that comprise victimology. The bottom line is, "Who was the victim and what was going on in his or her life at the time of the event." The best sources of information will be friends, family, associates, and neighbors, and that will be the initial focus of the investigation.[6]

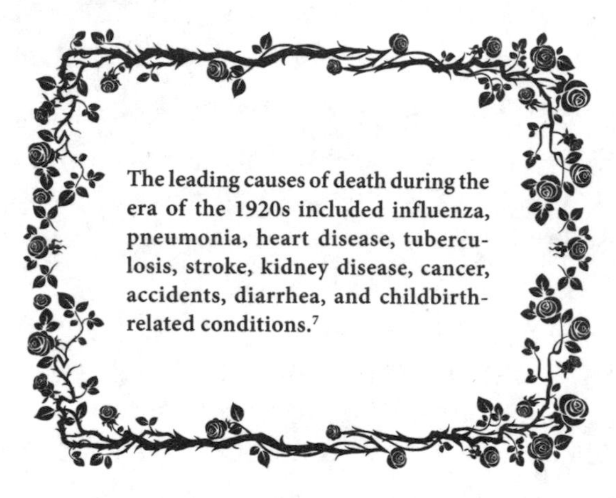

The leading causes of death during the era of the 1920s included influenza, pneumonia, heart disease, tuberculosis, stroke, kidney disease, cancer, accidents, diarrhea, and childbirth-related conditions.[7]

The first guess at the cause of death in *The Murder of Roger Ackroyd* is gastritis. What is it? Gastritis is the inflammation of the lining of the stomach and can be caused by a variety of factors including excessive alcohol use, the regular use of pain relievers, and bacterial infections. Gastritis can lead to stomach ulcers and bleeding but can be treated with medications like antibiotics, proton pump inhibitors, and acid blockers.[8]

The door is locked from the inside in *The Murder of Roger Ackroyd.* Keys date back all the way to ancient Babylon and Egypt in 704 B.C.E. when both keys and locks were made from wood. Iron and bronze were used in ancient Rome to create a skeleton key design, a cylindrical shaft with one single, thin rectangular bit, that was used for seventeen centuries![9] Flat keys became standard in the mid 1800s and utilized tumbler locks. Society has since progressed to electronic locks and key cards. What would Agatha Christie or her investigators think of this modern technology? And how would it play a role in her tales? A modern reimagining would be fun to watch or read.

Footprints on the windowsill reveal how the killer got in and escaped in *The Murder of Roger Ackroyd*. According to the Crime Museum, "Every person's foot has a unique set of ridges that make up a print unmatched by any other human being. As with fingerprints, the footprint's pattern is a unique characteristic that can pinpoint any one particular person. An actual footprint can be checked and matched to an existing print on record, such as one from a birth certificate."[10] Of course, most crime scene investigations may not include bare footprints, so shoe prints are studied to determine the size of the foot, the type of shoe worn, the suspect's height, and even weight.

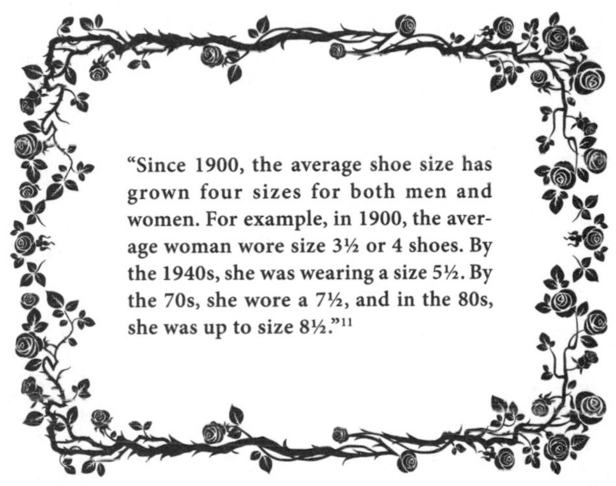

"Since 1900, the average shoe size has grown four sizes for both men and women. For example, in 1900, the average woman wore size 3½ or 4 shoes. By the 1940s, she was wearing a size 5½. By the 70s, she wore a 7½, and in the 80s, she was up to size 8½."[11]

The Murder of Roger Ackroyd remains one of the best examples of an unreliable narrator in mystery fiction, yet the science posited by Hercule Poirot is very much real.

CHAPTER THREE

The Seven Dials Mystery

Before we get into the fictional mystery of the Seven Dials secret society, we would be remiss not to unravel the very real mystery of Agatha Christie's 1926 disappearance. As outlined in the previous chapter, Christie met with a number of stressful life events around the publication of *The Murder of Roger Ackroyd*, namely the death of her mother and the infidelity of her husband, Colonel Archibald Christie. As thorough as Agatha Christie's several autobiographies are, she never mentions the strange events that began on a cold December night. After kissing her sleeping daughter goodnight, thirty-six-year-old Christie loaded a single briefcase into her Morris Cowley roadster. She left a note for her secretary that she would not return that Friday night. Christie was fond of driving so an overnight trip perhaps wasn't wholly unusual, but when she didn't return home the next day, her family and neighbors began to worry. Over that weekend, a hundred police officers combed the area of Berkshire, and that worry soon amplified into panic. It was especially palpable when her car was found in a troubling state as described by *The New York Times*:

> At eight o'clock yesterday morning the novelist's car was found abandoned near Guildford on the edge of a chalk pit, the front wheels actually overhanging the edge. The car evidently had run away, and only a thick hedge-growth prevented it from plunging into the pit. In the car were found articles of clothing and an attaché case containing papers.[1]

The article goes on to include a snippet from an interview with Colonel Christie in which he states that his wife was having a "nervous breakdown" with a counterpoint from an anonymous friend saying that Agatha Christie was "happy in her home life and devoted to her only child." It's important to note that the public did not know of Archie Christie's love affair at the time of Agatha's disappearance (although Agatha was more than aware).

As fans all over the world fretted that their favorite crime novelist could be the victim of harm, a letter was received by Agatha's brother-in-law. Several days after her disappearance, Archie's brother claimed to have received a note from her stating that she was in ill health and had gone to Yorkshire for a spa treatment. At this, the wide search for Christie was suspended as investigators researched the validity of the letter. Unconvinced that it was the truth, the police recommitted to finding her. They had the opinion that she had taken her own life, searching a local pond to no avail. As the days passed, thousands of people took part, as well as six bloodhounds, and Agatha Christie's own dog who had no talent in sniffing his owner's trail. Spiritualists even visited the chalk pit and held a séance in search of answers. A week had passed, and no one had come close to finding her. Rumors flew that she was alive; dressed in men's clothing in London, or wandering the moors of England in a state of catatonia. One tantalizing clue to motive was offered by a local:

> It is stated by one of Mrs. Christie's friends that the house in which she lived in Sunningdale was "getting on her nerves." It stands on a lonely lane, unlit at night, which has a reputation of being haunted. The lane has been the scene of a murder of a woman and the suicide of a man and its tragic associations were felt by Mrs. Christie. "If I do not leave Sunningdale soon, Sunningdale will be the end of me." She once said to a friend.[2]

A car crash at night, talk of hauntings, police combing for clues; it was as if Christie's inner, creative mind had come to life in rural England. One can imagine how Hercule Poirot would lead such a curious investigation of a missing, world-renowned novelist.

On December 15, eleven days after she slipped into the shadows, Agatha Christie was found in a Yorkshire spa, as she told her brother-in-law. She checked in under the surname "Neele," which the public would learn later was the name of her husband's mistress. Christie spoke only once of what had led to her week-and-half long disappearance to the *Daily Mail* in 1928 which was further explained in a more recent *Daily Mail* article:

> "There came into my mind the thought of driving into it (a nearby pond)," she said. "However, as my daughter was with me in the car, I dismissed the idea at once. That night I felt terribly miserable. I felt that I could go on no longer. I left home that night in a state of high nervous strain with the intention of doing something desperate." After driving around aimlessly, she stopped by the river at Maidenhead but realized that even if she were to throw herself into the water she was too good a swimmer to drown. Finally, she returned to Newlands Corner. "When I reached a point on the road which I thought was near the quarry, I turned the car off the road down the hill toward it," she said. "I left the wheel and let the car run. The car struck something with a jerk and pulled up suddenly. I was flung against the steering wheel, and my head hit something. Up to this moment I was Mrs. Christie." Feeling dazed and confused, she then caught a train from a station—most likely either Clandon or Guildford—to Waterloo. She crossed London and from King's Cross travelled to Harrogate, where she remained until she was reunited with her husband ten days later.[3]

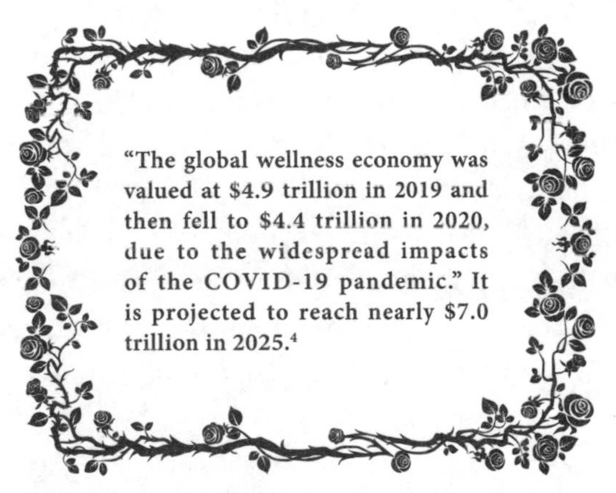

It's said that Agatha, on the steps of the spa, met her husband with a "stony stare" which would become more understandable to outsiders when she filed for divorce fifteen months later and Colonel Christie's affair came to light.

Nearly one hundred years later, we're still fascinated by what happened during those eleven days. Did Christie lose her memory? Was it a ploy to get Archie's attention? A break from reality for a person suffering with depression? Many have speculated, including author Kathleen Tynan, whose novel *The Summer Aeroplane* (1975) was adapted into the 1979 film *Agatha*, starring Vanessa Redgrave as Christie. The movie, with a touch of romance, is pure fiction, as is the British made-for-TV film *Agatha and the Truth of Murder* (2018), in which it is imagined that Christie was solving a murder with her deductive talents during that December of 1926. Most recently, the novel *The Christie Affair* (2022) combines fact and fiction from the perspective of Colonel Christie's mistress.

Amid the turmoil, Agatha Christie continued to write vivid stories of crime with her signature wry humor. *The Seven Dials Mystery* is more like her previous novel, *The Secret Adversary*, populated by young characters with a zest for adventure rather than the rich, middle-aged tea-sippers of *The Murder of Roger Ackroyd* and *The Mysterious Affair at Styles*. After the suspicious deaths of the young gentlemen Gerry Wade and Ronny Devereaux, one of our favorite Christie heroines, Eileen "Bundle" Brent, springs into action to uncover the truth behind what seems to be an international band of spies. Previously introduced in Christie's novel, *The Secret of Chimneys* (1925), Bundle is a spirited young adventurer with little interest in what society deems appropriate for a woman of her financial stature. She shows bravery when hiding in a cupboard to watch a meeting of the Seven Dials secret society and stops at nothing to find the culprits of her friends' murders.

It's interesting to note that the topic of marriage is spoken of frequently in *The Seven Dials Mystery*, including the unfortunate relationship between Lady Coote and her husband, Sir Oswald, who is often critical of her and has moved them into larger and more opulent homes without her approval:

> Lady Coote was not nearly so happy about it. She was a lonely woman . . . with a pack of housemaids, a butler like an archbishop, several footmen of imposing proportions, a bevy of scuttling

> kitchen and scullery maids, a terrifying foreign chef with a "temperament," and a housekeeper of immense proportions who alternatively creaked and rustled when she moved, Lady Coote was as one marooned on a desert island.[5]

One must wonder if this description of Lady Coote's existence was at all similar to Agatha Christie's, as her fame and funds grew and her relationship with her husband collapsed.

Despite the several negative outlooks of marriage throughout the book, Bundle has a charming romance with fellow amateur sleuth Bill Eversleigh and agrees to marry him at the conclusion of the novel. A happy ending for a pair who find out the truth of the Seven Dials society (it's not so bad after all) and unmask the true villains (who they thought were the friends in arms).

One other return character of note is Scotland Yard's Superintendent Battle, who also made his first appearance with Bundle in *The Secret of Chimneys,* and unlike Bundle goes on to appear in three more Christie novels. He appears as a guest at a dinner party in *Cards on the Table* (1926) with Poirot himself.

Gerry has a habit of oversleeping in *The Seven Dials Mystery*. What causes this? Oversleeping is officially defined as sleeping more than nine hours in a twenty-four-hour period, but surely we have all missed our alarms or hit snooze more than once. Two disorders that could contribute to oversleeping are hypersomnia and narcolepsy, which can be treated with medications and lifestyle changes. Can too much sleep be a bad thing? According to medical experts, oversleeping can be linked to heart disease, diabetes, and even an increased risk of death![6] Some of the factors that contribute to these conditions can often be linked to depression and low socioeconomic status, two things that may prevent people from seeking medical attention or having access to it. It's important to check with your doctor if you think you may be sleeping too much and to develop healthy sleep patterns like going to bed and waking up at the same time every day, avoiding screens and caffeine close to bedtime, and making your sleep environment as comfortable as possible.

What are some tips and tricks for waking up on time? Besides adjusting your lifestyle and sleeping quarters, consider physically moving your alarm far away from you. You'll be forced to get up and out of bed and hopefully stay awake. A 2018 study found that sleep fragmentation, repeatedly being woken up, then going back to sleep, has negative effects on pain tolerance and pain sensitivity.[7] It's better not to hit that snooze button! Next, let the sunlight into your bedroom, especially in the morning, and it can help regulate your circadian rhythms, which will improve your sleep. Last, consider setting important tasks or appointments for earlier in the day. Our brains are more likely to follow through and remember waking up for necessary things, like a doctor's appointment, rather than optional ones, like that workout you've been meaning to do.

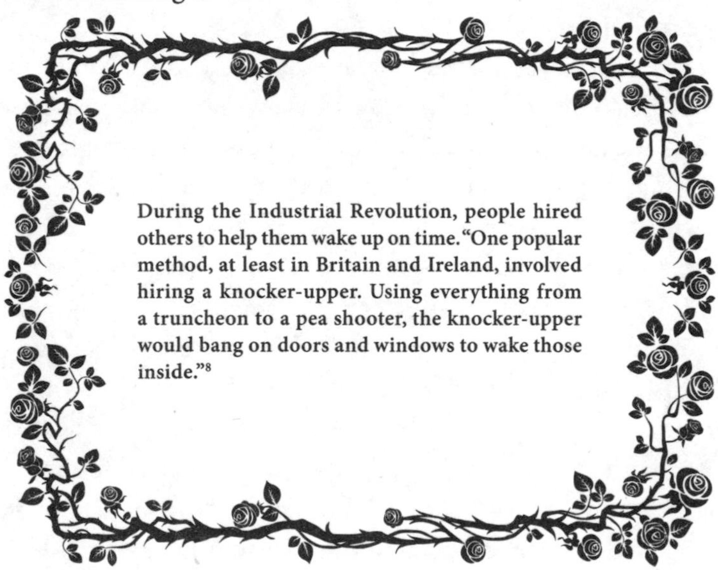

During the Industrial Revolution, people hired others to help them wake up on time. "One popular method, at least in Britain and Ireland, involved hiring a knocker-upper. Using everything from a truncheon to a pea shooter, the knocker-upper would bang on doors and windows to wake those inside."[8]

Gerry's determination of death in *The Seven Dials Mystery* is "death by misadventure." What exactly does this term mean? According to the *Handbook of Death and Dying*, it is a death attributed to an accident that occurred due to a risk that was taken voluntarily.

In contrast, when a cause of death is listed as an accident rather than a misadventure, this implies no unreasonable willful risk.

> Misadventure is a legally defined manner of death: a way by which an actual cause of death (trauma, exposure, etc.) was allowed to occur. Misadventure is a form of unnatural death, a category that also includes accidental death, suicide, and homicide.[9]

In 2019, child actor Mya-Lecia Naylor died of hanging at the age of sixteen and it was ruled death by misadventure. Although some clues were left that suggested she may have died by suicide, like low grades and being grounded, the assistant coroner in her case determined that she did not intend to end her own life, a sentiment her father agreed with.[10]

"Fire and arson investigators examine the physical attributes of a fire scene and identify and collect physical evidence from the scene. This evidence is then analyzed to help determine if the cause of the fire was accidental or deliberate."[11]

A charred, left-handed glove is found in the fireplace in *The Seven Dials Mystery*. What materials burn and which are more difficult to dispose of with fire? Flammable objects burn more easily, including wood, kerosene, and alcohol. Examples of nonflammable materials include helium, glass, and steel. Partial burned remains of evidence can be used by investigators at crime scenes to determine a variety of clues such as context, timing, and possible motive. In this case of this novel, teeth marks were found on the glove, and this helped explain who the killer was. How have teeth marks helped investigators in the past? Forensic dentistry helps identify individuals based on the properties of their teeth. "Looking at the location, orientation, presence or absence of teeth, and dental work, people can be matched to dental records or bite mark impressions for identification."[12] Although this sounds like a very modern practice, it's been around for centuries. According to the National Museum of Dentistry:

> The first known application of forensic odontology occurred during the rule of the Roman Empire with the case of Lollia Paulina. In 49 AD, at the request of Julia Agrippina, the wife of Emperor Claudius, Lollia Paulina, Agrippina's rival, was ordered to commit suicide. As proof that the act had been committed, Agrippina had a Roman soldier bring the head of Lollia Paulina for her to inspect. The story goes that due to the facial features being distorted from the act of suicide and decapitation, Agrippina could not immediately confirm the head did indeed belong to her rival. Then she remembered Lollia Paulina had unique teeth, which Agrippina examined the mouth for, resulting in the confirmation that her rival was indeed no more.[13]

In the United States, Paul Revere identified a man's remains by his teeth after he died in the Battle of Bunker Hill. Over one hundred years later, the first court case in the United States to utilize dental evidence took place in Boston, Massachusetts. Bite mark evidence used to identify suspects isn't necessarily 100 percent accurate, though, according to the California Innocence Project, because bite marks are often found on items of clothing or skin:

> Human skin is elastic; it swells, heals, and it can deform or warp a bite so that it does not align properly . . . Another problem with bite mark evidence is its similarity to other "sciences" such as fingerprint analysis and firearm analysis: they are subjective to the

> person evaluating the evidence. Different experts have found widely different results when looking at the same bite mark evidence.[14]

They go on to cite cases in which men were wrongfully convicted based on bite mark evidence who went on to be released from prison after DNA and other evidence proved them innocent.

The Seven Dials is a secret society. What is the history and prominence of such organizations? Secret societies have been around for centuries, and members of these clandestine groups include everyone from politicians to members of the working class. They were most popular during the eighteenth and nineteenth centuries and are described as:

> The incubators of democracy, modern science, and ecumenical religion. They elected their own leaders and drew up constitutions to govern their operations. It wasn't an accident that Voltaire, George Washington, and Ben Franklin were all active members. And just like today's networked radicals, much of their power was wrapped up in their ability to stay anonymous and keep their communications secret."[15]

Some notable ones include the Order of Elks, the Independent Order of Odd Fellows, and the Molly Maguires. Secret fraternal organizations include the Order of Skull and Bones, Freemasons, and the Illuminati.[16]

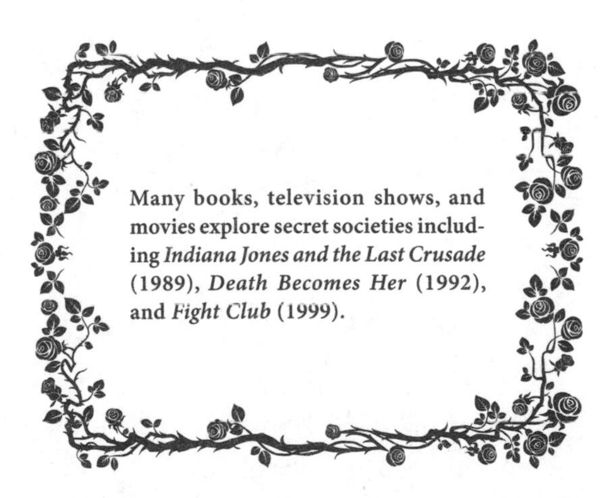

Many books, television shows, and movies explore secret societies including *Indiana Jones and the Last Crusade* (1989), *Death Becomes Her* (1992), and *Fight Club* (1999).

Although the secret details of Agatha Christie's disappearance may never be revealed, we know we can always count on her fictional mysteries to tie up loose ends in a delightful and satisfying way.

CHAPTER FOUR
The Sittaford Mystery

"It is generally understood in books," he said, "that the police are intent on having a victim and don't in the least care if that victim is innocent or not as long as they have enough evidence to convict him. That's not the truth, Miss Trefusis, it's only the guilty man we want."[1]

As the character of Inspector Narracott has just described, the issue at the heart of *The Sittaford Mystery* (or *Murder at Hazelwood* as it's known in the US) is that there is a suspect charged with the murder of Captain Trevelyan, yet he is clearly not the guilty party. How do the readers and characters know? Just as Agatha Christie makes fun of the tropes of mystery novels like above, she also employs them. It is simply too easy, too *boring*, that the deceased's nephew, who was the only one known to have visited his uncle that snowy night, is to blame. Arrested early in the novel, we can rest assured that the nephew, James Pearson, will be exonerated.

Without Hercule Poirot by chance visiting the rural towns of Sittaford and Exhampton, the mystery is left in the capable hands of Inspector Narracott, as well as the fiancé of the accused, Emily Trefusis, and journalist Charles Enderby. Unlike Poirot, Narracott himself is not a character who stands out from a lively cast of characters. He is clever, observant, and even once described as good looking, but he's really only a sidekick for the star of the book, Emily.

Emily is much in the same vein as "Bundle" Brent in Christie's previous *The Seven Dials Mystery*. She is headstrong, beautiful, and determined to solve the murder. Like Bundle, who locked herself in a cupboard for hours to find out the truth, Emily is bold. She convinces Charles Enderby to help her, with the use of her infectious charm, to the point that he falls madly in love with her. Too bad for Charles, as she is devoted to her imprisoned fiancé for the entirety of the novel. Despite no investigative experience, Emily questions witnesses, spies on possible suspects, and eventually, with little help from the men around her, names the correct murderer.

While Hercule Poirot and Miss Marple are Christie's most well-known investigators, we can't help but wonder what could've been if Emily or Bundle were given dozens of murders to solve. These forward-thinking young women of the era were given feminist ideals by Christie that shine through in their respective novels.

Published several years after Christie's own strange disappearance, *The Sittaford Mystery* is the first book of three to be dedicated to M.E.M.: Max Edgar Mallowan. Christie met him by chance. After a public divorce, Christie, now a forty-year-old single mother, spent time in therapy to rebuild her life. She wanted to leave dreary England behind for a touch of desert and sun and embarked on a trip to restore herself. At a dinner with friends, the author was convinced to take the *Orient Express* to Baghdad. She was also told to visit Ur, a city with ties to the Hebrew Bible in Mesopotamia, in order to survey the work of archeologists. Founded in the fourth millennium BCE, Ur, now known as Tall al-Muqayyar, Iraq, was rich with buried knowledge of centuries past:

> The first serious excavations at Ur were made after World War I by H. R. Hall of the British Museum, and as a result a joint expedition was formed by the British Museum and the University of Pennsylvania that carried on the excavations under Leonard Woolley's directorship from 1922 until 1934. Almost every period of the city's lifetime has been illustrated by the discoveries, and knowledge of Mesopotamian history has been greatly enlarged.[2]

It so happened that Leonard Woolley's wife was a huge fan of Christie, and they became fast friends when Christie visited the dig sites in 1928. While *The Sittaford Mystery* is strictly British, it was this first, and then subsequent, trips overseas that would inspire a great number of Agatha Christie's novels. Even riding the *Orient Express* to Baghdad obviously struck a creative chord! Though she had started out on a solo adventure, Christie was no longer alone when she left Ur and headed out on the train.

Leonard Woolley's archeological assistant, Max Mallowan, accompanied the author to Baghdad. At twenty-six, Mallowan was fourteen years Christie's junior. That did not stop them from falling in love amid the backdrop of ancient civilizations and beautiful desert landscapes. (Go, Agatha!)

When Christie had to cut her trip short, as she was summoned home to England because her daughter Rosalind was ill, Mallowan came with her. They married soon after and would remain together until death parted them forty-five years later.

The victim in *The Sittaford Mystery* is not so lucky to have died with the love of a companion. Captain Trevelyan, a lifelong bachelor, is described numerous times as being a "woman hater" or averse to women's company. He is grumpy and short with any woman he meets. This is teased as a possible motive but ends up being what those in the mystery world call a "red herring." (This term will come up again in the chapter for *And Then There Were None* on page 63.) This literary device refers to when information is given that seems vital, but it is only there to misdirect you from the truth. Like believing a woman killed the captain because of his misogynistic attitudes, when in truth, a man was at fault for totally different reasons. Researchers have traced where this fishy term came from:

> [We] now trace the figurative sense to the radical journalist William Cobbett, whose "Weekly Political Register" thundered in the years 1803–35 against the English political system he denigrated as the Old Corruption. He wrote a story, presumably fictional, in the issue of 14 February 1807 about how as a boy he had used a red herring as a decoy to deflect hounds chasing after a hare. He used the story as a metaphor to decry the press, which had allowed itself to be misled by false information about a supposed defeat of Napoleon.[3]

There is an escaped convict in *The Sittaford Mystery*. What are the real statistics of prisoners who are able to escape? According to Statista, "in 2019, 2,231 inmates escaped from state or federal prisons in the United

States.[4] This is a decrease from the previous year, when 2,351 prisoners escaped from facilities across the United States." In England, "there were [just] nine escapees in 2018/19, down from thirteen the previous year."[5]

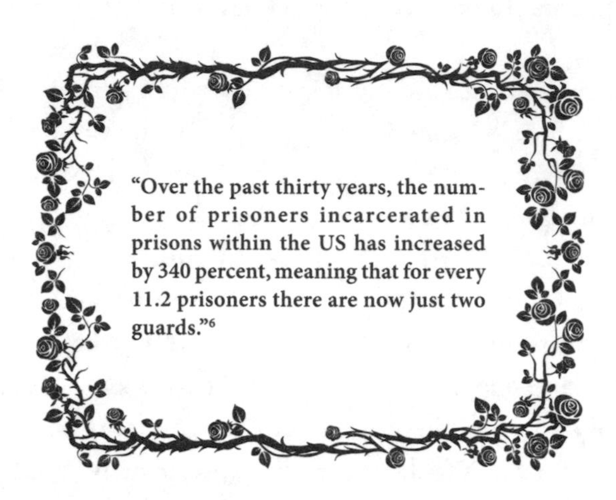

"Over the past thirty years, the number of prisoners incarcerated in prisons within the US has increased by 340 percent, meaning that for every 11.2 prisoners there are now just two guards."[6]

One of the scariest moments for me (Kelly) in *The Sittaford Mystery* is when a character loses her glasses. I felt this! I have always likened myself to Velma in *Scooby-Doo* (1969–1970) when she's blindly looking for her glasses on the ground while pursuing the monster of the week. What is the science and history of eyeglasses? Egyptians in the fifth century BCE experimented with eyesight corrections and by the first century CE the Roman Emperor Nero was using glass lenses to improve his vision. It wasn't until the thirteenth century that the first proper eyeglasses were invented. Benjamin Franklin is credited with the first bifocals and by the 1970s, contact lenses were made.[7] How do glasses work? Think of them as two rounded prisms put together. According to "The Science of Eyeglasses," "a prism is thicker at one end, and the light passing through is bent (refracted) toward the thickest portion. When a corrective lens (of the correct type of power) is placed in front of the eye, the focal point will be adjusted to compensate for the eye's inability to focus the incoming image onto the retina."[8] Depending on what type of correction you need for your eyes—nearsightedness, farsightedness, or astigmatism—your lens will change to accommodate to your specifications.

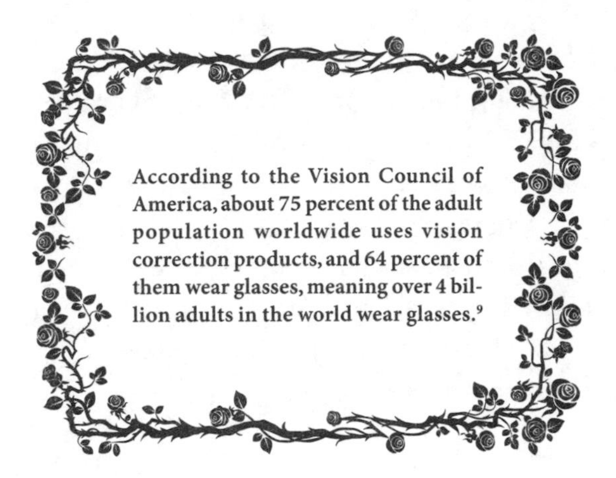

According to the Vision Council of America, about 75 percent of the adult population worldwide uses vision correction products, and 64 percent of them wear glasses, meaning over 4 billion adults in the world wear glasses.[9]

A round Ouija board is used in the film version of *The Sittaford Mystery*, and one finger is placed on a glass to navigate answers from the beyond. What is the history behind these boards? Did you know they began in America? According to *Smithsonian* magazine:

> Spiritualism worked for Americans: it was compatible with Christian dogma, meaning one could hold a séance on Saturday night and have no qualms about going to church the next day. It was an acceptable, even wholesome activity to contact spirits at séances, through automatic writing, or table turning parties, in which participants would place their hands on a small table and watch it begin to shake and rattle, while they all declared that they weren't moving it. The movement also offered solace in an era when the average lifespan was less than fifty: Women died in childbirth; children died of disease; and men died in war. Even Mary Todd Lincoln, wife of the venerable president, conducted séances in the White House after their eleven-year-old son died of a fever in 1862; during the Civil War, spiritualism gained adherents in droves, people desperate to connect with loved ones who'd gone away to war and never come home.[10]

Multiple people brave the harsh snowstorm to walk to the inn in *The Sittaford Mystery*. What are some tips if you need to brave a storm and go somewhere by foot? Experts at the National Weather Service recommend that you find shelter, stay dry, and try to cover all exposed body parts. If you are outside, build some sort of structure to protect yourself from the wind and build a fire if possible. Melt snow for drinking water and move your limbs to keep your blood circulating.[11] It sounds like you shouldn't go out, friends!

The "grin of death" (a smiling corpse) is mentioned in this novel. Does this actually happen? Unfortunately, yes. As seen in *The Black Phone* (2021), a sardonic grin on a face or a mask can be terrifying to behold. According to *Ancient Origins*:

> One drop of the poisonous hemlock water-dropwort was enough to completely incapacitate the target—the victim's muscles would grow taut, making it impossible to move, and the unusually uncomfortable smile would spread across the victim's face. Because of the "frozen" musculature, the face would remain like that. Meanwhile, the assassin would complete his or her murderous job.[12]

What do investigators use to help them determine a person's time of death? The first thing that crime scene investigators check is the temperature of the body. A body will lose one and a half degrees Fahrenheit per hour. An average body temperature is 98.6 degrees Fahrenheit, so a rectal temp is taken and the difference between the two, divided by one and a half, is the final number used to determine the approximate time of death.[13] Six hours after death rigor mortis sets in, and anytime from ten to fifty hours later, insects and other critters begin to feed on the body. Yikes!

Eyewitness testimony used to be the be-all and end-all to determine the guilt of a suspect in a court case, but science has proven that memory is not always reliable. All our memories are processed through our own filters and schemata, which put known information into categories in our brains. We view events through our own biases, filters, and experiences that influence how we handle memories. Sometimes the "misinformation effect" takes place when misleading questions contaminate a witness's memory.[14]

Sleepwalking occurs in *The Sittaford Mystery.* How common is this? According to *Smithsonian* magazine: "the first-ever large-scale survey of sleepwalking habits in American adults indicated that an estimated 3.6 percent of us—more than 8.4 million people—have had an episode of nocturnal wandering in the past year. This is much higher than researchers expected. Nearly 30 percent of respondents reported sleepwalking at some point in their lives."[15] Sleepwalking can range from briefly getting up to more complicated activities. The most frightening occurrence of sleepwalking to happen to us involved Meg's husband, Luke. He woke up, stared into a mirror, and was silent. Meg turned on a lamp to see what he was doing. "Turn off the light," he said as he stared back at her in the reflection of the mirror. She questioned him and he repeated it. Sounds like the beginning of a horror movie or mystery to us!

Multiple people have motives in *The Sittaford Mystery*, and that reminded us of a movie where all the characters could be suspects as well. We had the chance to talk to actor and director Josh Ruben about his hilarious horror comedy *Werewolves Within* (2021) and what it takes to set up a contemporary whodunit.

Kelly: **"When directing a mystery, how do you use cinematic elements to bring out the suspense of the story?"**

Josh Ruben: "I find that the simplest technical means to bring out suspense in film is how you utilize space and time. Allowing sequences to breathe, almost uncomfortably long, with smart use of camera movement in space will make or break suspension. John Carpenter is a genius at this, allowing us to 'live' with his heroes (and victims) in environments raked with shadows with minimal cuts, teasing our eyes with the potential that a killer might be lurking in the space. Sound design and score are also a biggie, but smart use of space with blocking and timing, that's the good stuff."

Meg: **"What clues did you leave for the audience, visually or otherwise, to hint about who the killer was in *Werewolves Within*?"**

Josh Ruben: "The fun of making a whodunit is scattering breadcrumbs and misleading the audience. I dialed each character's 'level of implication' meter up and down depending on the scene. Sometimes I'd have various characters—including the revealed killer—play each scene more or less suspiciously. We'd have fun dialing their behavior up and down to throw off the audience. I'd also block multiple 'layers' in each scene. Depending on where you were looking in any ensemble sequence, you might find two characters conversing suspiciously about doing something shady in a corner, while another slinks upstairs in soft focus, while another gazes curiously at another character who had no idea they were being watched within the scene. We got away with it because all characters were shady, to a degree. I made sure that if not all members of the cast needed to be on camera for the bigger ensemble scenes, we'd purposely exclude at least one character so we could potentially implicate them as a murder suspect. That was fun."

Kelly: **"Psychologists have studied how when people feel self-conscious, especially in heightened situations, they become more paranoid and their distrust of others is heightened. How do you think this paranoia or self-consciousness played a role in the group dynamics in *Werewolves Within*?"**

Josh Ruben: "We have a few scenes centered around this very notion. Paranoia is everything; the sense of distrust among characters. It was my job to remind my cast of where we were in the story, what happened in the scene previous, and what's coming; why X character feels Y about this or that. The cast had done their homework, so most of my job was to dial back or forward that 'self-conscious paranoia,' but that dynamic was everything in *Werewolves Within*. It was my job as the director to make sure the cast didn't 'get caught' overplaying it so they'd be believable enough for our audience to come along for the ride."

Meg: **"In the film, Dr. Jane Ellis is able to employ modern science to help solve the mystery of what's happening in the town of Beaverfield. How do you see science playing a role in horror movies?"**

Mary Shelley was influenced by the scientific discoveries of her time, which inspired her to write *Frankenstein* (1818).

Josh Ruben: "Science in horror is everything! My favorite horror films are when absolute horror is grounded in fact. That makes the supernatural or fantastical *possible.* When I'm reading a great genre fiction or screenplay with an otherworldly threat, I'm more inclined to sleep with the lights on if that threat is 'possible' given scientific boundaries; if the writer dresses up the fact that X + Y = Z every time, but that extra variable is a possibility—that's what gives us fans goosebumps. Storytellers who do the research and use science to spin tales are the most likely to make a lasting impact on their fans."

Kelly: **"How do you navigate finding the tone and balance in a horror comedy versus a straight comedy or straight horror movie?"**

Josh Ruben: "Pardon my sounding like an acting teacher, but this is the heart of finding tone: actors acting from a truthful, grounded place. If you're starring in a horror/comedy, whatever you do, *do not get caught trying to be funny.* If you're doing a straight comedy, you can go as far as you want to play or land a joke, but to get *caught* landing that joke dismantles the entire thing. In straight horror, of course, it's the combination of truth and composition. Guess it's safe to say the through line isn't just a killer cast with a preternatural awareness of their instrument; it's also your ability as a filmmaker to be a barometer for *how far is too far*; is that scream noticeably funny or chillingly real? Is the delivery of that line distractingly 'funny' or just distractingly bad acting? Both have just as horrid an outcome (not in a good way)."

Meg: **"Do you have a favorite Agatha Christie book/movie/play? If so, what do you like about it?"**

Josh Ruben: "Shamefully, I haven't read any Agatha Christie (*I know!*), but I can say my favorite part of any mystery is the reveal, not necessarily when the killer's identity is found, but when it dawns on our protagonist who the killer is, or who they've been all along, and gaining momentum toward the inevitable confrontation . . . few things are better."

We agree! Thank you to Josh for sharing his inside look into the making of *Werewolves Within*. If you haven't seen it yet, definitely go check it out.

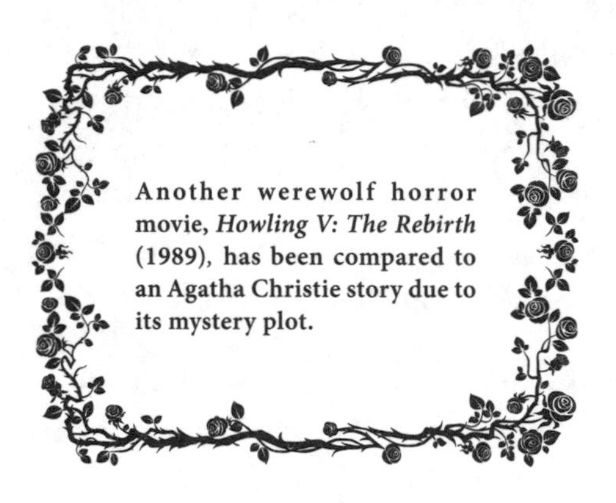

Another werewolf horror movie, *Howling V: The Rebirth* (1989), has been compared to an Agatha Christie story due to its mystery plot.

CHAPTER FIVE

Murder on the Orient Express

Perhaps the most well-known of Agatha Christie's oeuvre, the 1934 novel *Murder on the Orient Express* continues to shock modern readers with its gut-punch of a twist ending. In the same devious manner as *The Murder of Roger Ackroyd*, Christie plays upon her readers' expectation of a whodunit, only to playfully pull the rug out from under us. Perhaps it is this lasting thrill that inspired so many adaptations. These include two Hollywood films stacked with famous casts (in 1974 and 2017), as well as a five-part BBC radio play in 1992, an Audible recording in 2017, and popular TV versions (2001 and 2010).

Before we delve into the murderous fiction that Christie set upon the famous *Orient Express*, let us invite you on board to learn its history. The maiden voyage of the *Orient Express*, a train route developed by the Belgian company Compagnie Internationale des Wagons-Lits, traversed the European and Asian countryside before Agatha Christie was even born. The ambitious brainchild of George Nagelmackers, the *Orient Express* was the first of its kind: a luxury train that crossed continents. On October 4, 1883, the *Orient Express* set out on its first formal journey, with many journalists aboard to publicly marvel at the train's luxury and beauty. (Nagelmackers, a clever showman, even arranged to have shoddy, decaying old Pullman cars stand in contrast on the tracks adjacent to the *Express* as it left Paris's Gare de Strasbourg.) Aboard the train, the delighted passengers felt as though they'd entered one of Europe's finest hotels; they marveled at the intricate wooden paneling, deluxe leather armchairs, silk sheets, and wool blankets for the beds. The journey from Paris to Istanbul lasted a little over eighty hours.[2]

As popularity of the *Orient Express* burgeoned, so did its various routes, cutting through the mountainous terrain of the Alps, as well as bringing riders to European hot spots like Milan and Venice. It wasn't until 1977 that the last train from Paris to Istanbul pulled out of the Calais station, as the *Orient Express* fell victim to the rise of air travel and high-speed rail. But don't despair! Curiosity, thanks in part to the legacy of Agatha Christie's work, revived the *Orient Express*. You can now book a rail journey in a train that is very much in keeping with what Christie would've experienced in the early 1930s. Replete with art-deco finishes, the new *Orient Express* provides luxurious dining, well-appointed suites, and conductors dressed in antique flair. And even better? Its famous Paris-Istanbul route takes merely one night thanks to modern locomotives.

On her fateful trip to Ur in which Christie met her future husband, Max Mallowan, it was recommended by a friend that she take the *Orient Express* for the first time. The author recounts her eagerness to finally climb aboard the famous train in her autobiography:

> All my life I had wanted to go on the Orient Express. When I had travelled to France or Spain or Italy, the Orient Express had often been standing at Calais, and I had longed to climb up into it . . . Next morning I rushed around to Cook's, cancelled my tickets for the West Indies, and instead got tickets and reservations for a journey

> on the Simplon-Orient Express to Stamboul; from Stamboul to Damascus to Baghdad across the desert. I was wildly excited.[3]

Agatha Christie would continue to take the *Orient Express* throughout her life. An adventurous traveler, it was often her gateway to the Middle East, where husband Max was often on an archeological dig: "Next day I took the Orient Express on to Tell Kotchek, which was at that time the terminus of the Berlin-Baghdad railway. At Tell Kotchek there was more bad luck. They had had such bad weather that the track to Mosul was washed away in two places, and the wadis were up. I had to spend two days at the rest house." After the delay she finally made her way to Max in Mosul: "'Weren't you terribly worried,' I asked, 'when I didn't arrive three days ago?' 'Oh no,' said Max, 'it often happens.'"[4]

It is no wonder that Agatha Christie was inspired by her cross-country travels to write *Murder on the Orient Express*. Like the lavish mansions in many of her novels, the train cars, filled with finery and, of course, strangers, begged for a visit from Hercule Poirot.

With a cast of characters from all over the world, the novel plays upon the claustrophobic setting, especially when, much like Christie had experienced in real life, weather causes a delay. The snowstorm masterfully adds to the urgency when an American on the *Orient Express*, Mr. Ratchett, is found dead by the force of fifteen stab wounds. This means that someone on the train is responsible. Poor Hercule Poirot; every vacation turns into work! He sets to interviewing the passengers, doing what he does best: collecting clues with his nearly supernatural ability to connect the murderous dots. While the reader plays detective and puts themselves in Hercule Poirot's undoubtedly expensive and well-cared-for shoes, the ending of *Murder on the Orient Express* is not as simple as naming the culprit.

There are clues along the way which, as with many of Christie's stories, don't make sense until the whole picture is revealed. Like when Poirot examines the body:

> "Something strikes you as odd, does it not?" Dr. Constantine asked gently.

"Speak, my friend. There is something that puzzles you?"

"You are right," acknowledged Poirot.

"What is it?"

"You see the two wounds—here and here," he pointed. "They are deep, each cut must have severed blood vessels—and yet the edges do not gape. They have not bled as one would have expected."

"Which suggests?"

"That the man was already dead—some little time dead—when they were delivered. But that is surely absurd."[5]

Only the indomitable Hercule Poirot could explain the nature of stab wounds to a medical doctor! This odd clue, that Mr. Ratchett (who we later find out is actually named Cassetti) was stabbed after his death, is understandable when we reach the conclusion of the novel. In an unexpected turn, Poirot comes to reveal that *every* suspect stabbed Cassetti under the cloak of a dark and snowy night.

As *Murder on the Orient Express* unravels, we come to understand that mobster Cassetti is a victim that no one should mourn over. In his past, he kidnapped a three-year-old girl, causing a shock wave of grief and tragedy throughout her family, yet he still managed to evade the law. This injustice is too much for those affected, and, in the most brutal way, they take the law into their own hands.

Forever a man of honor, Hercule Poirot allows the truth of the murder to be covered up. He tells the police that a stranger surely climbed through

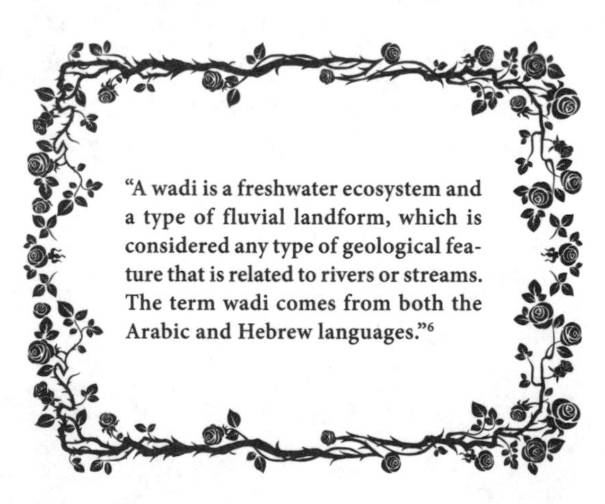

a window of the stopped train and stabbed Cassetti. The scoundrel, Poirot believed, had what was coming to him and, in the end, that justice was more important than the truth.

Poirot is particular about the size of his soft-boiled eggs in *Murder on the Orient Express*. How different can eggs possibly be? If you've only purchased eggs in a grocery store, you may be surprised to learn that eggs can vary widely in size, shape, and color. Store-bought eggs are purposely selected to be uniform in all ways to appeal to consumers but if you see eggs being laid on a farm, you'll notice they vary in thickness of the shell, may have markings like speckles or spots, and even vary in shape. These differences are all determined by the chicken's diet, circumstances such as stress while the egg is being laid, and environmental factors.[7]

Does Poirot have what we would now categorize as obsessive-compulsive disorder (OCD)? Actors who have played the famous detective, including Kenneth Branagh, drew from this condition to portray him on screen. When researching the role, Branagh said:

> I enjoyed finding the sort of obsessive-compulsive in him rather than the dandy and the prissy individual . . . I basically went through the books with a Sharpie and highlighted anything I thought was part of what might make mine sort of different. And I'd have a little checklist of the things I'd go back to and have your sort of identikit version of who he is, ultimately trying to leap off into something that you thought moment to moment could be spontaneous and real."[8]

David Suchet, who appeared as Poirot for decades in film and television, also believed the character to have OCD: "Well, there's no question he's obsessive-compulsive. Really so. When I'm him, every time I wash my hands out comes the lavender eau de toilette."[9] While Poirot likes order and method, are these traits enough to be diagnosed with OCD? Not necessarily. According to the Mayo Clinic, OCD typically includes repeated, unwanted thoughts, images, or urges that cause stress or anxiety. They can include a need for things to be orderly, a fear of things being dirty, and a difficulty dealing with uncertainty.[10] This condition can be treated with a combination of psychotherapy and medication and should be diagnosed by a doctor.

Cartography, the science and art of making maps, is used in *Murder on the Orient Express*. What is the history of this practice? Before paper, early cartography was done on clay tablets and cave walls. One of the earliest maps, dating back to 16,500 BCE, found on the cave walls in Lascaux, France, showed the night sky. The earliest map of the world found on a clay tablet was the Babylonian world map, created in 600 BCE. The Greek civilization helped develop the science of cartography by performing a "deep study of the size and shape of the earth and on habitable areas, climatic zones, and country positions."[11] During Roman times, "cartographers focused on practical uses: military and administrative needs."[12] Paper maps became the standard in Ancient Greece and were made more accurate with the inclusion of measurements like latitude and longitude. Maps continue to improve as technology does with the advent of satellite imagery and geographic information systems.

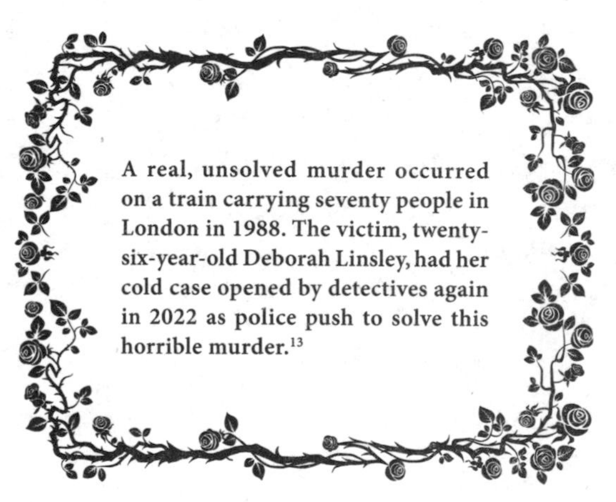

A real, unsolved murder occurred on a train carrying seventy people in London in 1988. The victim, twenty-six-year-old Deborah Linsley, had her cold case opened by detectives again in 2022 as police push to solve this horrible murder.[13]

Some characters in the movie are treated like celebrities by the press, and it almost feels like the photographers are paparazzi. What is the public's obsession with famous people and when did this phenomenon begin? The word "fame" has been used since Romans used it to share their opinions about people. According to Sharon Marcus, author of *The Drama of Celebrity* (2019):

> In what is now Europe, some of the earliest known celebrities were politicians, performers, and athletes. Several key figures in Plato's Symposium were Athenian celebrities . . . Medieval saints were also celebrities. But people who sought fame in classical and medieval times wanted to be remembered after their deaths. The goal of modern celebrity is to be renowned during one's lifetime.[14]

As technology continues to make bigger leaps and bounds, more people are in the public eye and experience fame on some level.

While a snowdrift stops the *Orient Express* in Christie's novel, an avalanche caused by lightning derails it in the 2017 film. What causes avalanches and how often do they occur? Snow can be weakened and set off downhill from a number of factors including precipitation, earthquakes, and melting snow. Ninety percent of avalanches are triggered by people who are partaking in activities like skiing, snowmobiling, and snowshoeing. In 2021, thirty-seven people died in avalanches in the United States,[15] while the deadliest avalanche in history killed an estimated 22,000 people in Peru in 1970![16] According to *National Geographic*, if you plan to be in an area that is susceptible to avalanches it's best to check forecasts for the current danger level, travel earlier in the day before snowpacks warm, and be aware of your surroundings. If you get caught in an avalanche, try to head at a forty-five-degree angle out of the path of the snow. If the snow catches up to you, reach for a tree or "swim" in the snow. "The human body is denser than avalanche debris and will sink quickly. As the slide slows, clear air space to breathe. Then punch a hand skyward."[17]

Coffee is laced with barbital in *Murder on the Orient Express*. What effect would this have on the body? Barbital was used as a sleep aid from 1903 until the 1950s due to its sedative and hypnotic effects. Because of its addictive qualities and users' tolerance levels, users could potentially overdose. Signs and symptoms of an overdose include shallow breathing, a weak but rapid pulse, respiratory failure, and possibly coma.[18]

In Bram Stoker's *Dracula* (1897), Count Dracula escapes from England to Varna by sea, while the group trying to destroy him travels to Paris on the *Orient Express*.

Mrs. Hubbard is stabbed in *Murder on the Orient Express*. Should the knife have been removed? According to the Red Cross, absolutely not! It's important to put pressure on or around the wound, call for medical help, and keep applying pressure until help arrives. Removing the object from the wound will only allow more blood to escape, making the situation worse. If the stab victim seems to be going into shock, it's important to elevate their legs higher than the rest of their body to keep blood flowing to their heart and brain.[19]

Poirot brings up the point that a single murder affects and touches multiple lives. The group of people affected work together in this story to commit murder. As we've written about in our book *The Science of Women in Horror* (2020), revenge doesn't necessarily bring peace. This group of people exact revenge as a way to cope with their collective and individual grief. How can we, instead, cope with grief in a healthy way? According to grief experts, let yourself feel some loneliness. It's okay to feel sad and sit with your feelings. Surround yourself with nourishing people who support you in healthy ways. Try not to set unrealistic expectations for yourself and instead take one day at a time. Make sure to get plenty of rest, both physically and mentally, to help you through this difficult time. Be prepared to feel your emotions when they come and talk to others about them if you're comfortable. Keeping up with physical activity, structure, and goals are important, and make sure to check in with professional help if you feel like you need it.[20]

And if all of the above suggestions don't work, please don't kill anybody on a train! Poirot may not be as gracious next time.

CHAPTER SIX

The Murder at the Vicarage

"I smiled. For all her fragile appearance, Miss Marple is capable of holding her own with any policeman or chief constable in existence."[1] In Christie's 1930 novel, *The Murder at the Vicarage*, narrator Reverend Clement gives the reader an apt description of his neighbor, Jane Marple. At the beginning of the novel's events, Miss Marple is often lumped in with the other elderly women of the quaint village of St. Mary Mead. She attends every church function and is often gossiping with her fellow "spinsters" like Miss Hartnell and Miss Wetherby. Despite her fragile appearance ("Miss Marple is a white-haired old lady with a gentle, appealing manner"),[2] the reader comes to discover that we can't underestimate such a dynamic and clever woman. As she herself puts it, "dear Vicar, you are so unworldly. I'm afraid that observing human nature for as long as I have done, one gets not to expect very much from it. I dare say idle tittle-tattle is very wrong and unkind, but it is often true isn't it?"[3] I (Meg) love how Miss Marple's advanced age is her superpower. She has become attuned to human nature, able to observe those living in St. Mary Mead with shrewd eyes.

> The brilliance of Christie's deployment of Miss Marple is that she does not turn away from the spinster stereotype. We all know it: old, unmarried women are lonely, nosy, and spend their days eavesdropping and passing judgment. And it is just her apparent superfluousness, the ease with which she may observe society, unnoticed and unimportant, that perfectly situates the spinster to pick up on clues and intrigue. Miss Marple's ambiguous position in domestic life, neither completely inside or outside the village families she observes, is where she gains her peculiar power to steer Christie's stories: as [Kathy] Mezei's mindblower of a sentence puts

> it, "[the spinster's] narrative function, in representing the dialectic between seeing and being seen, omniscience and invisibility, often mirrors the ambiguous and hidden role of the author/narrator in relation to his/her characters."[4]

Miss Marple's attention to detail comes in handy when a murder occurs at the vicarage next door. The victim, Colonel Protheroe, is known to be a rather bombastic blowhard with a healthy list of enemies. These include his wife and her lover, as well as his daughter, who is in need of his sizable inheritance. (Aren't we all?)

While Poirot would naturally place himself as the top investigator of such a crime, Miss Marple is far less accustomed to the spotlight. She tends to observe the suspects from afar, writing letters to the reverend when she has a theory or needs to work over a clue, thus avoiding the police. This trait shows itself at the end of the book, when Miss Marple allows the press to report that it was the detectives who solved the case, not her. She is content to stay on the periphery—something that the fastidious and egocentric Hercule Poirot would never stand for!

Twelve years after Miss Marple's debut in *The Murder at the Vicarage* (written when Agatha Christie was thirty-eight), she reappears in *The Body in the Library* (1942). It wasn't until Christie was in her sixties, closer in age to her character, that Miss Marple became a true staple of Christie's writing in the way the Belgian detective had.

"Her spinster detective was a development both of the Caroline Sheppard character she had enjoyed introducing in 1926 in *The Murder of Roger Ackroyd* and of the kind of old women she had known in her childhood in her surrogate grandmother's house in Ealing."[5] This "surrogate grandmother," Margaret West Miller, was known to Christie as "Auntie Grannie" and these two women, one real and one fictional, held a special trait in common according to Christie: "Though a cheerful person, she always expected the worst of everyone and everything and was, with almost frightening accuracy, usually proved right."[6]

The popularity of Miss Marple led to an impressive number of adaptations, including her own BBC series from 1984 to 1992, in which every Marple novel was made into an episode. In 1986, "The

Murder at the Vicarage," starring Joan Hickson as Miss Marple, was nominated for a BAFTA. In fact, Agatha Christie herself wanted Hickson to portray the famous spinster, as she saw the actress in a play forty years before Hickson took on the role, and wrote to her, saying "I hope one day you will play my dear Miss Marple."[7] Unfortunately, Christie died before she could see Hickson embody Marple, yet her blessing made it impossible for Hickson to turn down the role! The actress, age seventy-eight when she starred in the first episode in 1984, felt she wasn't right to play Miss Marple. "I thought I was the wrong shape, that Miss Marple would be much fluffier than me, much more wearing shawls and things," said Hickson. "But I was persuaded and now, well—I can only do it my way."[8]

The Murder at the Vicarage was adapted again in 2004. This time, Geraldine McEwan plays the elderly sleuth. I (Meg) watched this one thanks to BritBox and was surprised by how many changes were made to the original story. I did enjoy the addition of a romantic backstory for Miss Marple in which we learn she was a mistress of a slain soldier. How scandalous! While this was a fiction of the screenwriter, it makes contextual sense. As pointed out by *The Washington Post*'s Rhys Bowen:

> Marple actually represents a whole generation of women whose hopes of marriage were dashed by the loss of over a million young men on the battlefields of World War I. As a young woman of good family at that time, she was raised to make a good match, and not equipped for much else. She clearly has an excellent brain. In other times she might have gone on to university and had a lucrative profession. Instead, she has to content herself with her garden and good works around the parish. It's no wonder that she turns that excellent brain and sharp powers of observation to solving crimes.[9]

The article goes on to discuss the release of *Marple* (2022), a book in which twelve illustrious mystery authors of the twenty-first century (including Ruth Ware and Alyssa Cole) challenge themselves to write stories starring the one and only Miss Marple.

Agatha Christie was an unnecessarily harsh critic of *The Murder at the Vicarage* in her biography, yet her insight into the changes of post-World War II England is valuable:

> Reading *The Murder at the Vicarage* now, I am not so pleased with it as I was at the time. It has, I think, far too many characters, and too many subplots. But at any rate the *main* plot is sound. The village is as real to me as it could be—and indeed there are several villages remarkably like it, even these days. Little maids from orphanages, and well-trained servants on their way to higher things have faded away, but the daily women who have come to succeed them, are as real and human—though not, I must say, nearly as skilled as their predecessors.[10]

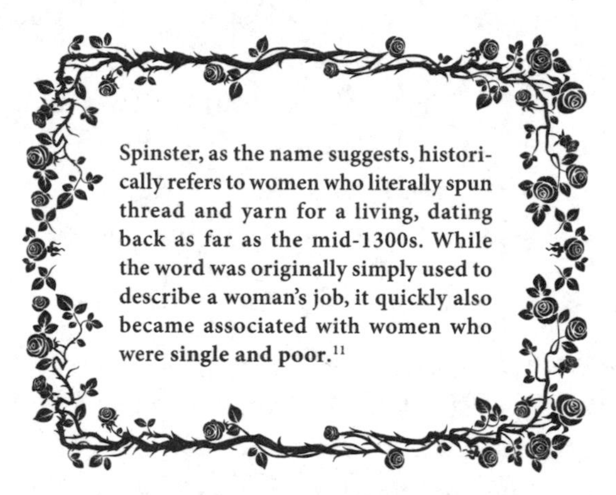

Spinster, as the name suggests, historically refers to women who literally spun thread and yarn for a living, dating back as far as the mid-1300s. While the word was originally simply used to describe a woman's job, it quickly also became associated with women who were single and poor.[11]

Miss Marple is a keen observer of goings-on in the neighborhood by noticing people's nonverbal communication. What can one's nonverbal behaviors tell us about them? Eye contact, or lack thereof, can reveal much about a person's mood, interest, and truth telling. Oculesics is the study of eye movement, and it has shown that various cultures feel differently about the length of direct eye contact. Some cultures find it rude while others require it to prove respect. Proxemics, or the use of personal space and distance in nonverbal communication, can reveal how comfortable you feel with someone and how protective you are over your personal space. Coined by Edward T. Hall in 1963, proxemics generally deals with four distances: intimate, personal, social, and public spaces.[12] Intimate space is anywhere from zero to eighteen inches. If you are comfortable with someone within this space, you would be okay sitting right up next to them on a bench. This is usually reserved for the closest relationships in your life, including your significant other and family members. Personal space is anywhere from eighteen inches to four feet and is most common among family members and close friends. Social distance is from four to twelve feet and is typically seen among coworkers and classmates. Public distance is anywhere more than twelve feet and is typical at events such as concerts or lectures. Other nonverbal traits Miss Marple may observe include tone of voice, gestures, and posture.

The detective in *The Murder at the Vicarage* is ageist toward Miss Marple. He assumes she will have a hard time remembering details and even speaks louder as he asks if she can hear him all right. Ageism is discrimination based on someone's age and applying stereotypical traits to them without getting to know them. To combat ageism in our society, it's important to become aware of it and confront one's own stereotypes head-on. Agatha Christie attracted many female readers by introducing this female supersleuth to the world. "Are there any secrets left in St. Mary Mead?" the vicar asks. A townsperson responds, "Lots yet, I hope!" Miss Marple, indeed, went on to appear in many more Christie stories.

Time frames come into play quite a bit during the investigation of the murder. How does chronemics play a role in our lives? Chronemics is the study of time and how it relates to human communication. Some people are very monochronic in their approach to daily life and take on one task at a time, prefer to have things scheduled, and engage in little to no small talk. Polychronic people and cultures are less concerned with a rigid schedule, will often multitask, and will take part in small talk more often. Chronemics also includes how people view time whether it be past, present, or future. Past-oriented people

The social clock is an idea that varies by culture, which determines the right time in life for certain milestones to occur such as college, marriage, and having children.

will focus on the past, whether for good or bad, and often bring up old memories and stories. Present-oriented people live in the here and now, not focusing too much on the future. According to the American Psychological Association, future-oriented thinkers tend to save money, make healthy choices, and be the most successful due to their focus on planning ahead.[13]

Miss Marple recognizes a French cigarette by its scent in *The Murder at the Vicarage*. How does our sense of smell help us in our daily lives? Smell can help us protect ourselves by alerting us to rotten food, someone we are attracted to, and even potential danger. Some people are considered "super smellers," and they can detect and are more sensitive to scents. This condition, hyperosmia, is rare and appears in people who have certain medical conditions including pregnancy, Lyme disease, and migraines.

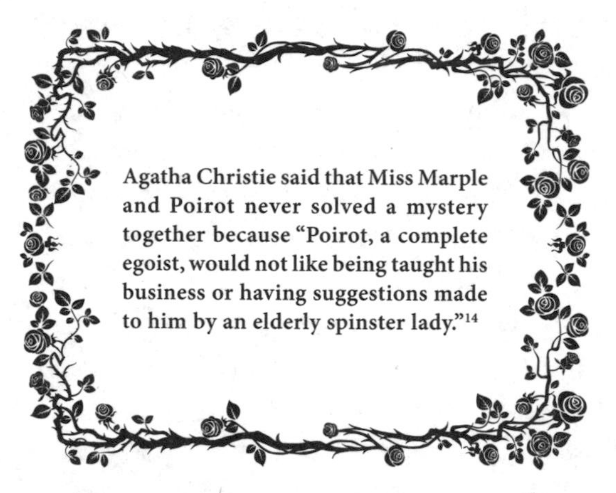

Agatha Christie said that Miss Marple and Poirot never solved a mystery together because "Poirot, a complete egoist, would not like being taught his business or having suggestions made to him by an elderly spinster lady."[14]

Handwriting analysis is used in *The Murder at the Vicarage* and has been used for over four hundred years. In 1622, Camillo Baldi began observing how one's handwriting could reveal their "nature and qualities."[15] Although graphologists still assess an individual's handwriting to determine personality traits, a more forensic approach is used by investigators to match handwriting samples to solve crimes. The first step is to analyze the writing and compare it with any other known document that was written by the suspect. Second, it's important to determine if the writing is original or if it is copied. Third, see if the writing is distorted or natural in its quality. Finally, see if any variations occur that could point to inconsistencies such as letter formations, the quality of the lines, alignment of the writing itself, and arrangement such as spacing.[16] By comparing and evaluating these findings, it can be determined whether the samples are written by the same person.

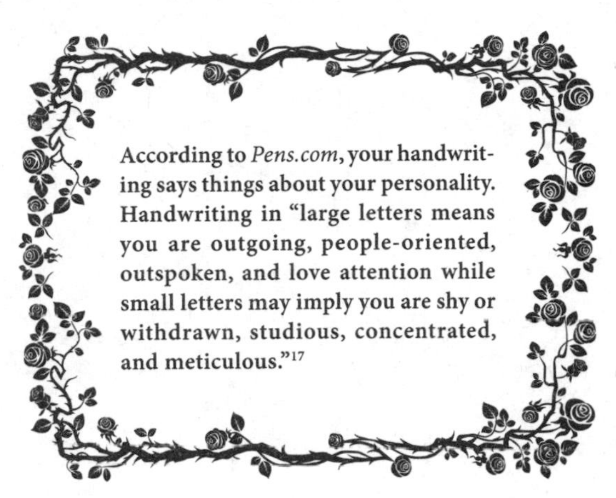

According to *Pens.com*, your handwriting says things about your personality. Handwriting in "large letters means you are outgoing, people-oriented, outspoken, and love attention while small letters may imply you are shy or withdrawn, studious, concentrated, and meticulous."[17]

Even though Agatha Christie was a bit harsh in her criticism of *The Murder at the Vicarage*, we believe it is an essential Miss Marple mystery.

CHAPTER SEVEN

Death on the Nile

At only seventeen years old, recent high school grad Agatha Christie spent a winter in Cairo with her mother, escaping the doldrums of an English winter for heat and palm trees. "The city left a distinct impression: Christie would base her first novel, the unpublished *Snow Upon the Desert,* on her experiences in the Gezirah Hotel in Cairo. Christie sent her manuscript to several publishers, and it was turned down, perhaps as she had expected given her inexperience at that time."[1] *Snow Upon the Desert* was a romantic novel, written shortly after teenaged Christie returned home to England in 1908. It remains unpublished, as the gossip is that it is "quite amateurish."[2] Knowing she worked tirelessly to hone her writing craft after the disappointment of *Snow Upon the Desert* makes her an even more formidable Queen of Crime!

We researched the Gezirah Hotel and, unsurprisingly, found that it sat right upon the Nile River. What once was a palace of a Khedive (an Egyptian viceroy under Turkish rule) named Isma'il Pasha, it had been converted into a hotel at the turn of the twentieth century. The hotel, designed to resemble Versailles, was given a sizable addition by Marriott in the 1980s and now runs as the Cairo Marriott Hotel & Omar Khayyam Casino, one of the largest hotels in the Middle East. Perhaps after a ride

Christie discovered she could use her diluted face cream to properly clean ivories found on an excavation in Nimrod.[3]

on the restored *Orient Express*, a die-hard Christie fan would enjoy a stay—especially in a room that overlooks the Nile!

While Christie was obviously inspired by Cairo's exotic architecture, it wasn't until she was in her forties that she was drawn to excavate the ancient Egyptian ruins. As Egyptologist Jun Yi Wong asserts, digging alongside her husband and his colleagues may have boosted her creativity:

> Christie regularly accompanied Mallowan on his excavations in the Middle East. She immensely enjoyed the lifestyle in the field, typically spending the start of the excavation season writing, then dedicating herself fully to fieldwork when the excavation intensified. Christie soon became highly knowledgeable, despite considering herself an outsider to the discipline—a "happily amused onlooker," as friend and archaeologist Jacquetta Hawkes described. If her work in the field had any bearing on her creativity, it was probably a constructive one: During this period, Christie regularly produced two or three books annually, many of which count among her best works.[4]

It's no wonder that Agatha Christie found her fascinating desert surroundings the ideal setting for a novel. Considered one of her masterpieces, *Death on the Nile* (1937) is yet another labyrinthine and enjoyable mystery, with Hercule Poirot at its center. Once again, the Belgian detective is simply trying to enjoy his holiday, this time in Egypt, when murder strikes!

A love triangle is at the heart of *Death on the Nile*, as we come to find out that extremely rich Linnet Doyle has "stolen" her best friend's (Jacqueline De Bellefort) fiancé, the penniless but undeniably handsome Simon Doyle. The newlyweds, Linnet and Simon, are desperate to enjoy their honeymoon in Egypt, yet Jacqueline is determined to follow them, popping up to remind them of their betrayal. One has to wonder if Christie held some sympathy for young Jacqueline after the disintegration of her own first marriage in which a woman "stole" her husband. Like any sensible socialite, Linnet asks Hercule Poirot for help in getting Jacqueline to abandon her plans of torment. Yet, despite Poirot's urging for Jacqueline to stop for her own benefit, she continues, placing herself on the SS *Karnak* with the detective, the Doyles, and a cast of suspicious characters who may want Linnet Doyle dead.

Like many of Christie's stories, *Death on the Nile* has been adapted numerous times for TV and film. My (Meg) personal favorite is 2004's *Poirot* installment in which Emily Blunt plays the first victim and man stealer Linnet Doyle. Just as Christie described in the novel, Blunt's Doyle is frustratingly beautiful with a mean streak. This, naturally, means she is the perfect Christie victim: someone with multiple enemies who the reader doesn't miss very much! *Death on the Nile* was also adapted into a computer game in 2007 in which players enact the role of Hercule Poirot, as well as a French graphic novel, also in 2007, adapted by François Rivière and illustrated by Solidor. Most recently, a 2022 episode of *Chucky* (2021–) plays on the familiar tropes that Christie so eloquently developed and is even titled "Death on Denial."

The "Agatha Christie Suite" on the SS *Sudan*, nearly one hundred years old, is available for your cruise of the Nile. It provides a starboard view, as well as luxurious gold furnishings.[5]

In 1936, Christie sent Hercule Poirot to Iraq for *Murder in Mesopotamia*, where he grew accustomed to detecting foul play among the archeological ruins. He references this in *Death on the Nile*:

> Once I went professionally to an archaeological expedition—and I learnt something there. In the course of an excavation, when something comes up out of the ground, everything is cleared away very carefully all around it. You take away the loose earth, and

> you scrape here and there with a knife until finally your object is there, all alone, ready to be drawn and photographed with no extraneous matter confusing it. This is what I have been seeking to do—clear away the extraneous matter so that we can see the truth.[6]

Perhaps this is why Christie found such interest in her husband's work. Finding clues within layers of sediment is similar to the methodical nature of solving a crime. As Christie gained experience in archeology, she began to take on more roles, including hiring laborers, ordering supplies, and eventually cataloging and restoring artifacts.

One trip in particular with husband Max Mallowan solidified her writing of *Death on the Nile*:

> In 1933, the Mallowans stopped in Egypt on their way to an archaeological dig, boarding the SS *Sudan* for a Nile River cruise. Built in 1885 for Egypt's royal family and converted into a cruise ship in 1921, the luxury vessel—still in operation today, complete with a suite where Christie allegedly stayed—took visitors to the Nile cataracts, Luxor and Aswan. Most passengers were members of the European elite.[7]

In contrast to contemporary novels like *Death on the Nile* and *Murder in Mesopotamia*, Agatha Christie challenged herself to fully embody her surroundings by writing a mystery set in the twelfth dynasty (1991–1802 BCE). "*Death Comes as the End* (1944) is the only one of Christie's novels not to be set in the twentieth century and not to feature any European characters. The death of a priest's concubine sets off a series of murders within the family and, as in Christie's more familiar 20th-century whodunits, the scene is soon littered with bodies."[8]

Christie read the preserved letters of a priest living in the twelfth dynasty, Heqanakhte, to draw upon daily Ancient Egyptian life. These letters weren't sensational; rather, accounts of pretty typical day-to-day affairs steeped in landowning and farm business. Though there were a few glimmers of drama, like the mention of some upset when the priest's second wife joined the family. Armed with the simplest of plots, Christie did what she does best: wove a murderous tale into *Death Comes as the End* that, despite its unconventional period and setting, resonated with her readers.

Christie further stretched her writing prowess with *Come, Tell Me How You Live* in 1946, an autobiographical travelogue in which she shares her experiences of digs across the Middle East in the 1930s that had come to an abrupt end because of World War II. It is described as "giving an intimate insight into their [Agatha and Max's] travels together as well as many of the inspirations which fed into her novels and plays."[9]

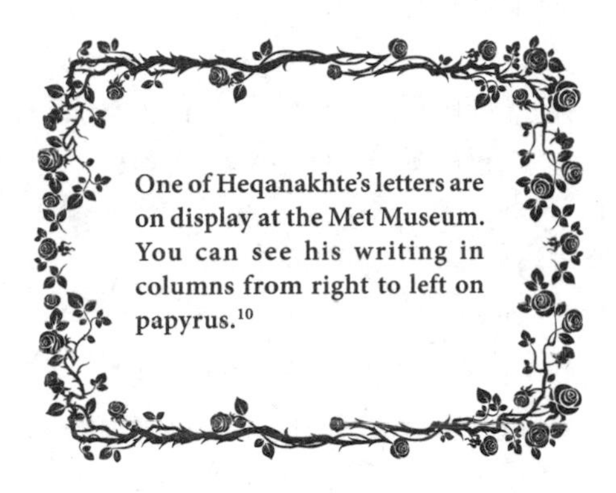

One of Heqanakhte's letters are on display at the Met Museum. You can see his writing in columns from right to left on papyrus.[10]

Hercule Poirot observes bird flight patterns to predict weather in the beginning of the film version of *Death on the Nile* (2022). What can these teach us about meteorology? According to *The Farmer's Almanac*, low-flying birds can signal rain while high-flying birds are a sign of fair weather.[11] Barometric pressure, which birds are sensitive to, allows them to predict desirable flying conditions. For example, cold fronts are signaled by low barometric pressure and birds will tend to avoid flight in this type of weather. Researchers also believe some birds hear low-frequency noises that can help them prepare for thunderstorms before humans can.[12] By observing birds, we can hypothesize not only about upcoming weather but also wind direction, speed, and patterns.

The pyramids of Egypt play a role in *Death on the Nile* as a backdrop. These iconic landmarks were created between 2550 and 2490 BCE and stood for over four thousand years as the tallest structure on Earth at 481 feet. Scientists have studied how the pyramids could have been built and have come up with various theories. The granite used to construct them came from a quarry nearly five hundred miles away and was transported to the site by ships on the Nile. Through experimentation and research, experts believe the stones were moved by sleds that were able to be pulled more easily by wetting the sand in front of them, reducing friction. Physicists discovered this technique by studying a painting on a tomb from 1900 BCE that illustrated it. They tested the theory and proved that it worked. Other theories are still being tested of how the stones could have been raised, including the use of pulleys or ramps.[13]

A theme in *Death on the Nile* is a woman, Jacqueline, being driven by jealousy and revenge. What is the psychology behind this? Jealousy can range from feelings of humiliation to anger and sadness to fear. Jacqueline de Bellefort is feeling all these emotions as her fiancé's heart is stolen by her best friend. Although evolutionary psychologists regard jealousy as an important emotion to signal relationship preservation, it's not always healthy.[14] Jealousy can be a sign of low self-esteem, feeling possessive, or having a fear of abandonment. Talking about the feelings out loud to the other person, reducing negative self-talk, and seeking therapy can all reduce jealous tendencies.

The newlywed couple, Simon and Linnet, are on a honeymoon with a group of their closest friends in *Death on the Nile*. According to a study of fifteen thousand couples who got married in 2021, 68 percent of them planned to take a honeymoon with an average cost of $4,600.[15] We're curious what the cost of Simon and Linnet's honeymoon trip would be today!

What is the history of the tradition of honeymoons with friends or on your own? The practice of what we typically consider a honeymoon began in nineteenth-century England, where couples would travel, sometimes with a group of their friends and family, to visit those who couldn't attend the wedding. By the late 1870s, it became more common for couples to travel by themselves for a honeymoon trip. The word itself has had various etymological theories. One theory, which has not been verified, purports that the term comes from the amount of honey mead that was given to couples after they married to serve as a moon's worth of aphrodisiac to ease the couple into intimacy and ultimately conceiving a baby. Another reference from 1818 refers to the honeymoon as "the first month after marriage, when there is nothing but tenderness and pleasure."[16] This "honeymoon phase" of a relationship can last longer than a month according to psychologist Rachel Levenson. She suggests continuing to spend time with your partner doing things you enjoy and communicating openly.[17]

According to a survey by Expedia, some of the most popular honeymoon destinations in the world are Paris, France; Los Angeles, California; and Rome, Italy.[18]

Poirot, after having a glass of champagne, is feeling a bit seasick. What happens to our bodies when we feel this way? Seasickness, or motion sickness, is considered a physiological vertigo and affects humans and animals when they are placed in unfamiliar situations like riding in cars, on boats, or other activities. Symptoms include nausea, vomiting, pallor, and cold sweating.[19] How can you treat this? If riding in a vehicle, it's important to keep your head motions as stable as possible and to look forward. Therefore, those who suffer from motion sickness often do better driving or sitting in the passenger seat. According to the CDC, some drugs may be helpful in preventing or treating the symptoms of motion sickness, including antihistamines and anticholinergics.[20]

Several gunshots are analyzed in *Death on the Nile*. One gunshot proved that the gun had be fired directly to the temple, while another proved that the gun had been muzzled and silenced with a scarf. How are bullet holes and bullet wounds investigated by professionals? Forensic ballistics experts study the numerous pieces of evidence that firearms leave behind, including marks on the bullet and cartridge case, while other forensic analysis can prove the distance of the shooter, the cause and manner of any injuries, and the type of weapon used.[21] With Poirot's investigative background, he would be aware of many of these traits.

Related to this, bloodstain pattern analysis is used by Poirot to determine who killed the victim. Although we may see this science portrayed in today's TV crime procedurals, it is not a new field:

> In 1895, Dr. Eduard Piotrowski of the Institute of Forensic Medicine in Krakow, Poland, published a paper on bloodstain pattern analysis titled "Concerning Origin, Shape, Direction, and Distribution of Bloodstains following Head Wounds Caused by Blows." Dr. Piotrowski performed experiments using live rabbits, white paper, and a variety of instruments including rocks, hammers, and hatchets to better understand how bloodstains are created and what information could be gleaned from their study.[22]

For analysts today, this is an interpretation of bloodstains that can help form opinions about what happened at a crime scene. Bloodstain pattern analysis uses physics, biology, and math to analyze where the blood came from, how the victim was positioned, from what direction the victim was wounded, and movements made after blood was shed.[23]

Could a bullet go through two people and kill them as in the end of *Death on the Nile*? It *is* possible, according to ballistic experts. A .22 caliber bullet can travel approximately a mile and a half with no obstacles, while a 9 mm bullet could go as far as three miles.[24] The force, then, of a bullet could go through two people at close proximity and kill them both.

The tragic love story in *Death on the Nile* is one reason why this novel is still popular today, as well as the ancient setting truly beloved by Christie herself.

CHAPTER EIGHT

And Then There Were None

Nearly thirty years ago, before Netflix and Amazon Prime gave us mountains of content, I (Meg) had my collection of VHS tapes recorded from TV. My favorites were *Ghostbusters II* (1989), *Clue* (1985), and a rather scratchy copy of *Ten Little Indians* (1965). I have a vivid memory of watching the black and white British film, immediately rewinding it, and starting it again from the beginning. Though I was the only fourth grader with a penchant for British murder mysteries, this didn't stop me from discussing *Ten Little Indians* with all who would listen. I even declared a crush on the one-named Fabian who played Mike Raven in the film, counterpart to the novel's Tony Marston.

I discovered *And Then There Were None* (1939) in the sixth grade. It was my first Agatha Christie novel that led not only to my love of her writing, but also to the writing of this book! *And Then There Were None* is the original American title, as *Ten Little Niggers* and then *Ten Little Indians* were the British titles. You can probably guess why both titles have been changed with the progression of time. The title of the famous poem that the killer in the book uses as a sort of guidebook in killing, "Ten Little Niggers," has also been changed to "Ten Little Soldiers." It reads as follows:

> Ten little Soldier Boys went out to dine; One choked his little self and then there were nine.
>
> Nine little Soldier Boys sat up very late; One overslept himself and then there were eight.
>
> Eight little Soldier Boys traveling in Devon; One said he'd stay there and then there were seven.
>
> Seven little Soldier Boys chopping up sticks; One chopped himself in halves and then there were six.

Six little Soldier Boys playing with a hive; A bumblebee stung one and then there were five.

Five little Soldier Boys going in for law; One got in Chancery and then there were four.

Four little Soldier Boys going out to sea; A red herring swallowed one and then there were three.

Three little Soldier Boys walking in the zoo; A big bear hugged one and then there were two.

Two little Soldier Boys sitting in the sun; One got frizzled up and then there was one.

One little Soldier Boy left all alone; He went out and hanged himself and then there were none.[1]

This poem was not invented by Christie, and in fact made a small appearance in the Sherlock Holmes–led novel, *A Study in Scarlet* (1887). It's a familiarly macabre minstrel's song that dates back to the 1860s. When the book was published on the cusp of the 1940s, the N-word was already considered an aggressively racist term in the US, though it was still used in British society. This necessitated the title change from the start.

***And Then There Were None* is Agatha Christie's bestselling book, and the sixth of all time! It's estimated to have sold one-hundred-million copies.[2]**

And Then There Were None's famous trope of strangers trapped with a murderer among them has become one of my most beloved plots in mystery and horror. I'm clearly not alone, as this trope has appeared frequently, from films like *House on Haunted Hill* (1959) to board games like Clue (invented in 1943). This also explains the international appeal of *And Then There Were None*, as it is continually adapted in films with languages such as Tamil, Tagalog, and Russian. It even has the dubious honor of inspiring the "adult film" *Ten Little Maidens* (1985)!

A few years after the release of *And Then There Were None*, Christie found a diversion from World War II in the world of theater. "Agatha was increasingly involved in the theater, adapting her works (notably *And Then There Were None*) to the stage, attending rehearsals and first nights."[3]

Unlike popular Christie novels like *The Murder of Roger Ackroyd* and *Death on the Nile*, this book is stuffed with multiple killings by a mysterious figure. Today, they would be known as a mass murderer. "Mass murderers kill many people, typically at the same time in a single location. With some exceptions, many mass murders end with the death of the perpetrators, either by self-infliction or by law enforcement."[4] At the end of the book, the guilty party does kill himself with a complicated pulley system in order to make it look as though he has been shot in the head by someone else.

The setup of *And Then There Were None* is criminal perfection: ten strangers are summoned to a remote island where they are confronted by the playing of a record. This voice recording lists how each person is responsible for the death of one or more people. As the story unravels, we hear the differing opinions on these deaths. Some, like religious spinster Emily Brent, deny any wrongdoing, while others, like General MacArthur, admit to their devious deeds. No matter their level of personal guilt, the murderer known as "U N Owen" or "Unknown" takes on the mantle of judge, jury, and executioner. Using the poem "Ten Little Soldiers" as inspiration, he kills the strangers one at a time. For example, the line "Seven little Soldier Boys chopping up sticks; One chopped himself in halves and then there were six" is affixed to the death of Mr. Rogers. He is killed with an axe to the head as he is chopping wood for the morning fire.

One of the most surprising inclusions in *And Then There Were None* is the mention of an 1892 murder that we detail in our book, *The Science of Serial Killers* (2021). When the men are debating whether women are capable of murder, ex-inspector Blore points out the true case:

> There was a case in America. Old gentleman and his wife—both killed with an ax. Middle of the morning. Nobody in the house but the daughter and the maid. Maid, it was proved, couldn't have done it. Daughter was a respectable middle-aged spinster. Seemed incredible. So incredible that they acquitted her. But they never found any other explanation.[5]

This obvious mention of Lizzie Borden's trial, without use of her name, seems to convince the others that a woman could indeed wield an axe to kill Mr. Rogers, as well as have the ability to kill all the others. It appears that Agatha Christie and I share a fascination with the Borden murders, as she references it again in Miss Marple's last case, *Sleeping Murder* (1976).

Christie ruminates on her most popular novel in her autobiography:

> I had written the book . . . because it was so difficult to do that the idea had fascinated me. Ten people had to die without it becoming ridiculous or the murderer being obvious. I wrote the book after a tremendous amount of planning, and I was pleased with what I had made of it. It was clear, straightforward, baffling, and yet had a perfectly reasonable explanation; in fact it had to have an epilogue in order to explain it. It was well received and reviewed, but the person who was really pleased with it was myself, for I knew better than any critic how difficult it had been.[6]

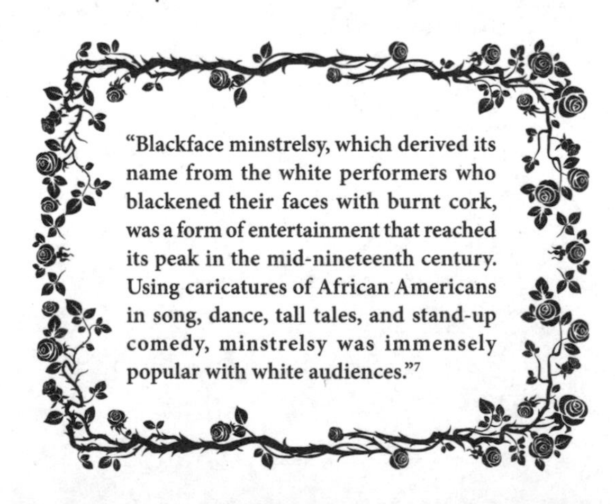

"Blackface minstrelsy, which derived its name from the white performers who blackened their faces with burnt cork, was a form of entertainment that reached its peak in the mid-nineteenth century. Using caricatures of African Americans in song, dance, tall tales, and stand-up comedy, minstrelsy was immensely popular with white audiences."[7]

The character of Anthony dies after choking on poisoned whiskey in *And Then There Were None*. How common is choking? It is most common among the elderly and very young children. There were nearly five thousand deaths by choking in the United States in 2020, which, considering the odds, is more common than dying in a plane crash![8] If you or someone else is choking, encourage coughing to clear the object. If one is not able to cough, perform the Heimlich maneuver, which involves standing behind the victim, placing your fist against their abdomen above the navel, and thrusting inward and upward with quick jerks.[9]

The character of Ethel dies in her sleep from an overdose of chloral hydrate in *And Then There Were None*. What is the history of this drug? First discovered in 1832, the sedative properties of this drug made it a popular choice to treat insomnia. It is still used as a sedative before minor medical and dental procedures. Many notable people have become addicted to chloral hydrate over the years including Hank Williams,[10] Marilyn Monroe,[11] and Anna Nicole Smith,[12] who all died from overdosing on it.

The character of Mr. Blore is crushed and killed by a marble clock shaped like a bear. How often does a falling object kill someone? According to *The New York Times*, it's quite rare. Six hundred and eighty people in the United States die each year in this way, or about two people per million.[13] Most often, these accidents are happening on construction sites with unsecured tools, equipment, or building materials. What factors contribute to how dangerous these situations can be? First, it depends on

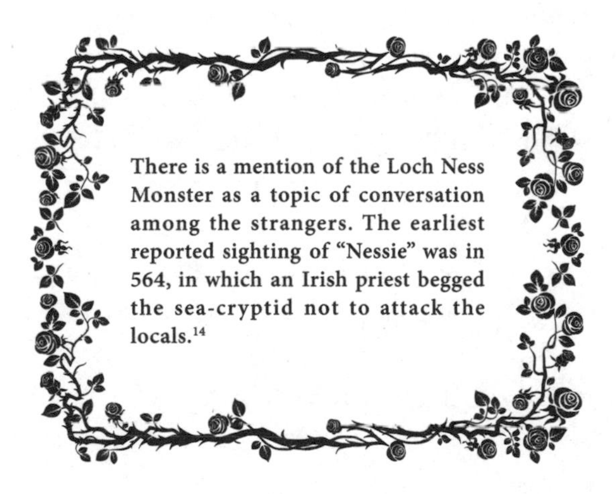

There is a mention of the Loch Ness Monster as a topic of conversation among the strangers. The earliest reported sighting of "Nessie" was in 564, in which an Irish priest begged the sea-cryptid not to attack the locals.[14]

the size and weight of the object that is falling. The bigger and heavier it is, the more potential for damage. The height from which it's dropped also plays a part. The higher up an object is dropped, the faster it travels and can cause fatal injuries.

The character of Vera is overcome by the strong smell of the sea and hangs herself to fulfill the final line of the song in *And Then There Were None*. How can smell affect us? According to psychology professor Rachel S. Herz:

> In olfaction, the process can be understood as follows: a novel odor is experienced in the context of an unconditioned stimulus, such as surgical procedure in a hospital, which elicits an unconditioned emotional response, such as anxiety. The odor then becomes a conditioned stimulus for that hospital experience and acquires the ability to elicit the conditioned response of anxiety when encountered in the future. This mechanism explains both how odors come to be liked or disliked, as well as how they can elicit emotions and moods.[15]

This can explain why we may have a strong reaction to an ex-partner's cologne or perfume, experience nostalgia when we smell our grandmother's cookie recipe baking, or feel sick to our stomachs if we get a whiff of an alcohol that didn't sit well with us the night before! Brands and stores understand this link and have taken to having signature scents pumped into their stores or products to create a memorable, pleasurable experience. Dawn Goldworm, internationally recognized olfactive expert, says "smell is the only fully developed sense a fetus has in the womb, and it's the one that is the most developed in a child through the age of around ten when sight takes over. And because smell and emotion are stored as one memory, childhood tends to be the period in which you create the basis for smells you will like and hate for the rest of your life."[16] This definitely made us rethink the scents we love and the scents we avoid in our own lives!

We had the amazing opportunity to interview chemist Dr. Kathryn Harkup, author of *A is for Arsenic: The Poisons of Agatha Christie*

(2017), about her own journey and research into the world of Christie's novels.

Kelly: **"Tell us about your journey getting into science and chemistry!"**

Dr. Kathryn Harkup: "I always enjoyed science at school, thanks to some excellent teachers. Of the many options within science, I decided to pursue chemistry at university. For me, chemistry combines theoretical and practical particularly well. I liked the idea of going into a lab and being able to make something no one has made before. Even after four years studying it, I was still hooked and carried on with postgraduate studies. I haven't been in a lab for decades. What I missed initially was the problem-solving aspect of research. I think that is why I always like Agatha Christie's books, for the puzzles she set. Now I try to solve problems like writing books and analyzing Christie's use of poisons."

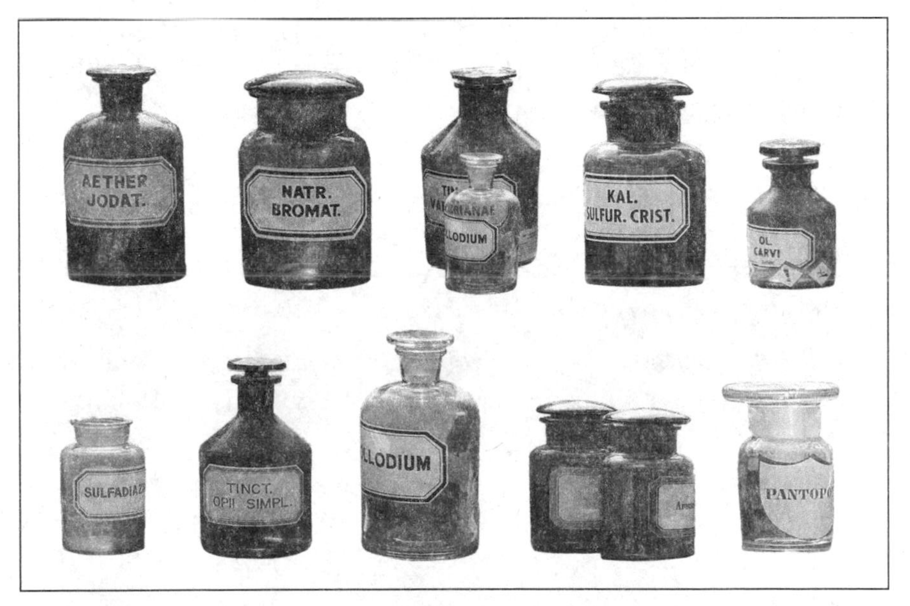

Agatha Christie's knowledge of chemistry was rooted in her volunteer work as a nurse during World War I. "This is a time when all drugs, pills, tonics, cream, etc. were made up by hand."[17]

Meg: **"Your book *A is for Arsenic: The Poisons of Agatha Christie* explores the science that Agatha Christie uses in her stories. How accurate did you find her to be with her use of poisons?"**

Dr. Kathryn Harkup: "Christie is impressively accurate in her use of poisons for fictional murders. Her descriptions of symptoms are second to none. However, her use of forensic techniques is scarce. Her real strength was in creative choices of poison and ingenious methods of administration. She may stretch credibility on occasion but, despite writing fiction, she almost never invents anything for the benefit of her plots. You really have to search for scientific errors in her work. If you find them, they are relatively minor."

Kelly: **"What was the most surprising thing you learned when researching Agatha Christie for your book?"**

Dr. Kathryn Harkup: "Christie's creativity and variety in the poisons she selected for her stories. What really stood out for me was how brilliantly those poisons are woven into her plots. Symptoms, accessibility, and medical histories all play an important part in giving clues and red herrings to the detectives, as well as the reader. Her poisons are so well described and the effects they have are so integrated into the narrative, they are almost characters in themselves."

Meg: **"Why were poison deaths so popular in the late nineteenth and early twentieth centuries? Has there been a change?"**

Dr. Kathryn Harkup: "There were a lot of notorious poisoning cases in the nineteenth century, probably because toxic substances were much easier to get hold of and forensic detection was in its infancy. Today there are three main factors making it less likely someone will be murdered by poison. First, changes in the law mean it is much more difficult to obtain substances that are lethal in small amounts. Improved medical care means a victim of poisoning is also more likely to survive. Finally, forensic science has made incredible advances, so poisoners are much less likely to get away with it."

Kelly: **"Is the idea that poison is a 'woman's weapon' appropriate?"**

Dr. Kathryn Harkup: "I don't think so. Yes . . . women are slightly more likely to use poison than a man to murder someone, but the percentages are small. Knives and blunt objects are far more likely to be used by either sex. Men are ten times more likely to commit murder than women, so there are far more male poisoners than female. I think the perception comes from the media. Women who kill have always attracted more press attention than men, and poisoners also get more press coverage. Combined, women poisoners get a lot of publicity, and it gives the impression that poison is a 'woman's weapon.'"

We loved hearing about Dr. Harkup's experience in chemistry and science and can't wait to check out her other books soon.

CHAPTER NINE

Murder in Retrospect

At the beginning of *Murder in Retrospect* (1942), also published as *Five Little Pigs*, newly engaged Carla Lemarchant comes to Poirot with a strange request. Sixteen years earlier, her mother, Caroline Crale, was convicted of killing her father, Amyas Crale. Although she was only six years old at the time, Carla can't believe that her mother was responsible.

> Listen, M. Poirot, there are some things that children know quite well. I can remember my mother—a patchy sort of remembrance, of course, but I remember quite well the *sort* of person she was. She didn't tell lies—kind lies. If a thing was going to hurt she always told you so. Dentists, or thorns in your finger—all that sort of thing. Truth was a—a natural impulse to her. I wasn't, I don't think, specially fond of her—but I trusted her.[1]

What struck us about this description of a mother long dead was the complicated nature of Carla's relationship with Caroline. It wasn't simply "I loved her, she didn't do it." It held more nuance, as real mother-daughter relationships do. This led us to want to know more about Rosalind, Christie's only child, and how their relationship may have influenced her characters.

Rosalind Christie, born in 1919, grew up amid her mother's fame and success. She was only seven when Christie went missing for those unaccounted days in 1926, as well as during the subsequent divorce from her father, Archibald Christie. Because of Agatha Christie's wealth and propensity for travel, Rosalind attended the posh girls' boarding school, Benenden, in Kent, continuing her studies in France and Switzerland. Benenden, first established as a school in the early 1920s, is still accepting applications for girls aged eleven to eighteen today!

Motherhood was a complexity in Christie's life, as the strict ideals of the Victorian age were diminishing, yet they still lingered. Christie was expected to balance her love for her daughter and her husband, causing no jealousy, while running a house and becoming the bestselling crime novelist of all time! Biographer Gillian Gill recounts the entrance of Rosalind into her mother's sphere:

> Archie's fears about a child rival were quite unrealized, since the baby turned out to be a girl and a small replica of her father in both good looks and temperament. Agatha's reaction to having a daughter was somewhat more ambiguous. Whereas both Archie and Clara (Christie's mother) were delighted with the birth of a girl child, Agatha herself wanted a son. The experience of giving birth seems also to have revived that uneasy sense of observing herself playing a part in her own life which Christie reports feeling from early childhood. In the autobiographical account of her daughter's birth that she gives in *Unfinished Portrait* (1934), Christie confesses that she felt quite unreal in her new role of young mother.[2]

Unfinished Portrait is a Christie novel, one of six written under the pen name Mary Westmacott, that is believed to be semiautobiographical, as Gill suggests. It holds clues as to how Christie might have interpreted her relationship with teenaged Rosalind whose counterpart is named Judy in the novel: "I don't know whether I've failed with Judy or succeeded. I don't know whether she loves me or doesn't love me. I've given her material things. I haven't been able to give her the other things—the things that matter to me—because she doesn't want them."[3] This exasperation of a mother with her teenager is nothing new, and it becomes clear that as Rosalind grows into adulthood, her relationship with her mother seemingly improves.

Rosalind's birth father was not a big part of her life as she grew older, so Christie's second husband, Max Mallowan, stepped in to raise her. "He introduced his stepdaughter to archaeology and the Near East. These expeditions, a London season, and other travel occupied Rosalind, who was tall and handsome, with dark hair and a high color, until her marriage in 1941, to Hubert Prichard, a regular soldier in the Royal Welsh Fusiliers."[4] The Prichards welcomed a son, Mathew (Agatha Christie's only grandchild), in 1943, shortly before tragedy struck. In 1944, Hubert died at the battle of Normandy. Now a widow, Rosalind and her one-year-old moved to the Greenway Estate, one of Christie's homes, located in Devon upon the Dart River. Agatha Christie described first seeing the home in 1938: "So we went over to Greenway, and very beautiful the house and grounds were. A white Georgian house of about 1780 or '90, with woods sweeping down to the Dart below, and a lot of fine shrubs and trees—the ideal house, a dream house."[5]

Such a dream, in fact, that Christie depicted the rural oasis of Greenway in her novels, such as the main house of the Crales on the Dart River in *Murder in Retrospect.* (Talk about full circle!) Rosalind continued to live at Greenway for the rest of her life, joined by her second husband, Anthony Hicks, in 1949. Rosalind and Anthony enjoyed a long life together, passing away in 2004 and 2005, respectively, a mere six months apart.

After the death of Christie in 1978, Rosalind took on the heavy mantle of her mother's legacy. She became president of the Agatha Christie Society in 1993, fastidiously cultivating the films, TV shows, and plays adapted from Christie's works. Rosalind approved of actors David Suchet (Hercule Poirot) and Joan Hickson (Jane Marple), even naming them co-vice presidents of the society.

Mathew Prichard continues the family business that his grandmother began and his mother continued. He allowed Greenway Estate to be sold to the National Trust, which means you can add it to your list of Agatha Christie tourist stops! The house and gardens are open to the public. At AgathaChristie.com, Mathew shares remembrances of his grandmother, or as he calls her, "Nima," in videos released every month.

An intriguing mother-daughter relationship is one appeal of *Murder in Retrospect*. Another is the novel approach of Poirot solving what we'd call a "cold case" today. Usually, Poirot is able to see the body and crime scene, perhaps even examine the victim before rigor mortis sets in. But Amyas Crale was poisoned sixteen years earlier, creating a unique challenge for our favorite Belgian detective. Poirot, confident as ever, is not worried about the case's limitations: "'One does not, you know, employ merely the muscles. I do not need to bend and measure the footprints and pick up cigarette ends and examine the bent blades of grass. It is enough for me to sit back in my chair and *think*. It is this' —he tapped his egg-shaped head— 'this, that functions!'"[6]

Technically, this murder is not a traditional cold case, as it had been considered solved and therefore closed. "A case becomes 'cold' when all probative investigative leads available to the primary investigators are exhausted and the case remains open and unsolved after a period of three years."[7] On the request of Carla Lemarchant, Hercule reopens the case of his own accord, interviewing the five people who were present at the famous artist's murder. Simply by using his "egg-shaped" head, Poirot is able to piece together Caroline's innocence from the witness testimony. In a rather unordinary ending, the perpetrator, jilted lover Elsa Dittisham, gets away with her cold-blooded deed. Sort of. It's clear that Elsa has evaded court justice, though she has condemned herself to a joyless life: "I didn't understand that I was killing *myself*, not him. Afterwards I saw her [Caroline] caught in a trap—and that was no good either. I couldn't hurt her—she didn't care—she escaped from it all—half the time she wasn't there. She and Amyas both escaped—they went somewhere where I couldn't get at them. But they didn't die. I died."[8]

Like in *Murder on the Orient Express*, Agatha Christie reminds us, for different reasons, that being tried by a judge is not always the most satisfying ending. Elsa Dittisham may be beautiful and rich, but she is dead inside. To someone as adventurous and creative as Christie, this would be a fate worse than prison, or even the hangman's noose.

Caroline Crale was wrongly convicted of murdering her husband in *Murder in Retrospect*. How often are people falsely accused? To learn more about this important topic, we spoke with Sara Jones, executive director of the Great North Innocence Project.

Kelly: **"Tell us how you got involved in this work and this organization specifically."**

Sara Jones: "My awareness of criminal law and matters of justice began in childhood, growing up with a father who served for over two decades as Minnesota State Public Defender and taught criminal and constitutional law. As a result, I had an early foundation that I didn't recognize until later. Much later, after shifting my career from law into nonprofit development, I was invited to serve on the board of the Council on Crime and Justice. There, I learned more about mass incarceration, the collateral consequences and ripple effects of a criminal record, flaws in our system, and the importance of research and policy in improving the criminal legal system. When this position opened up a few years ago, I felt like my personal and professional history called me to this work to try to help people wronged by our legal system. My father's modeling of service and commitment to justice had far more of an influence that I realized as a younger person."

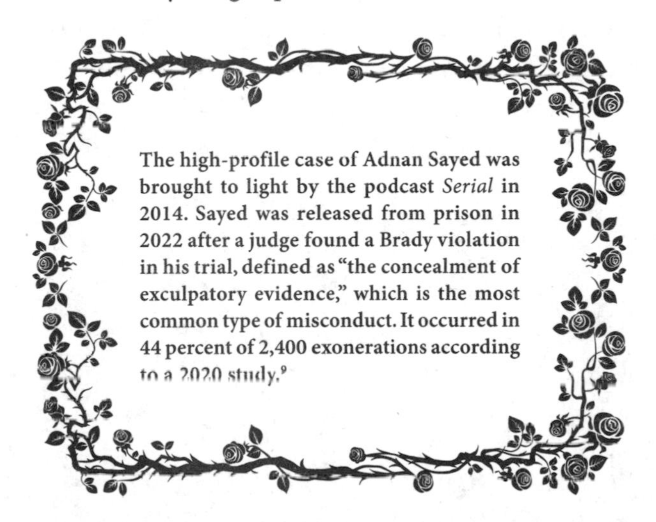

The high-profile case of Adnan Sayed was brought to light by the podcast *Serial* in 2014. Sayed was released from prison in 2022 after a judge found a Brady violation in his trial, defined as "the concealment of exculpatory evidence," which is the most common type of misconduct. It occurred in 44 percent of 2,400 exonerations according to a 2020 study.[9]

Meg: **"How have you seen the Innocence Project impact people's lives?"**

Sara Jones: "The Great North Innocence Project, and all of the organizations in the Innocence Network, make an enormous impact in not only the lives of those people who are freed after a wrongful conviction, but also the lives of their families and communities. The work we do also impacts all of us who count on the criminal-legal system to advance justice, uphold truth, and keep us safe. Innocence organizations give hope to people who feel lost and forgotten. It can restore some fullness to their lives even if we can't restore all of what could have been. For instance, our client Javon Davis, who was wrongfully convicted of a shooting in Minneapolis, was separated from his partner and six children while he was incarcerated. Now that he's free, he is able to be a fully present part of his children's lives. He helps to organize youth basketball leagues in Minneapolis and never misses one of his kids' dance recitals or sporting events. However, his family lost out on his love and support and financial contributions while he was wrongfully incarcerated, and he's still awaiting compensation for the state's mistake. As you can imagine, all of our clients who are incarcerated wrongly for any length of time experience isolation from family and community, loss of income, challenges to their identity and integrity, and trauma. Their families experience the same, and by the time an individual approaches an innocence organization for help, they've often exhausted all of their other options and lost time and money in the process, all through no

Prisons in Northern Europe, through a more supportive system, see lower rates of violence and recidivism.[10]

fault of their own. Advocating for these individuals and families and letting them know that they have an advocate working on their behalf without additional cost provides hope and a potential pathway to real justice."

Kelly: **"I can only imagine what they must go through!"**

Sara Jones: "Although the journey certainly doesn't end when a person walks out of the prison doors to freedom, it is the start of a new opportunity for a rich, free life. Additionally, our organization seeks to make changes in the criminal-legal system and educate current and future criminal-legal professionals, jurors, and community members about the criminal-legal system and fair processes so as to prevent wrongful convictions from happening in the first place, improving the criminal-legal system for all of us. We know many of the reasons that people have been wrongfully convicted, and the great news is that there are precautions, tools, and best practices that we can put in place to make sure those mistakes don't happen anymore. For example, we know that eyewitness misidentification is a leading cause of an innocent person being convicted of a crime. There are reforms that law enforcement officials and attorneys can easily put into place to increase the accuracy of eyewitness identifications, ensuring not only that an innocent person isn't convicted but that the person who actually committed a crime is held responsible. Things as simple as ensuring a lineup consists of individuals who closely match the eyewitness's description of the perpetrator, having the person who facilitates a lineup be unaware of who (if any) of the individuals in the lineup is the suspected perpetrator, instructing a witness that they're not required to select a person as part of a lineup, and asking a witness to state their level of confidence in their choice can greatly reduce errors in eyewitness identification and require very few resources."

Kelly: **"What impacts have the fields of science played in solving cases compared to decades ago?"**

Sara Jones: "Science is a tool that investigators can use to identify a perpetrator, exclude potential suspects, link crime scenes, and uncover additional details about a crime that can lead to a successful arrest and prosecution of a true perpetrator. DNA is the classic example. As DNA evidence has developed over time, we've been able to harness the powerful science to identify perpetrators of violent crimes decades after they've been committed. DNA was also the original fuel behind Innocence Project founders Barry Scheck and Peter Neufeld's work exonerating innocent people. With newer developments in genetic testing and genetic genealogy, investigators are now beginning to be able to identify perpetrators through close familial DNA matches and building family trees to narrow down suspect pools. It's powerful. However, not all 'science' is created equal. True scientific methodology includes ongoing testing of theories and updating our understanding when a particular method is disproven. The challenge legally, however, is that when a particular type of science is introduced into court and is used as evidence to convict a person during trial, that creates a precedent by which that methodology can be used in future legal proceedings. This doesn't necessarily change when the scientific community learns that a particular method of science is actually inaccurate or problematic. It can also be difficult to convey to judges, law enforcement professionals, lawyers, and juries (most of whom aren't scientists) why a method we once believed to be foolproof is actually junk science.

"Since 1989, of documented exonerations (which is not inclusive of every instance of a wrongful conviction), 23 percent of the wrongful convictions were, at least in part, the result of faulty forensic science. A great example, and an area where we've had cases of innocent people being wrongfully convicted, is in the diagnosis of abusive head trauma, formerly known as 'shaken baby syndrome.' When AHT/SBS first emerged as a diagnosis for infant deaths in the late 1970s, the theory was that if a specific triad of symptoms—bleeding on the brain, bleeding behind the retinas, and brain swelling—presented in a young infant, it was the result of violent shaking by whomever had been taking care of the child when the symptoms started. However, medical understanding of AHT/SBS has significantly evolved since then. Specifically, doctors have found that there are many possible causes of these three symptoms cooccurring that are unrelated to abuse, including physical trauma that can occur naturally during childbirth, an accidental fall several days prior to the symptoms' onset, a variety of diseases, and genetic abnormalities. Unfortunately, the earlier belief that the cause of 'the triad' was exclusively abuse by a caregiver has resulted in the wrongful conviction of many parents, relatives, and childcare providers, including one of our own clients Michael Hansen (who has been exonerated and freed). Other well-documented examples of 'forensic science' that have been debunked, and yet still show up in court proceedings and are linked to wrongful convictions, include blood spatter analysis, hair and fiber analysis, and bite mark analysis."

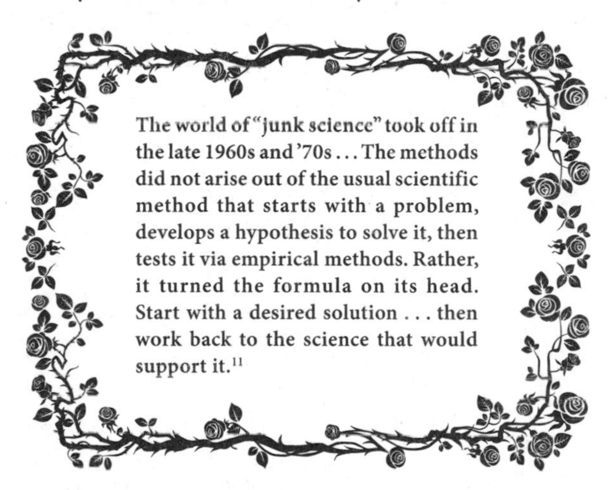

The world of "junk science" took off in the late 1960s and '70s . . . The methods did not arise out of the usual scientific method that starts with a problem, develops a hypothesis to solve it, then tests it via empirical methods. Rather, it turned the formula on its head. Start with a desired solution . . . then work back to the science that would support it.[11]

Meg: **"How can we, as regular citizens, help this cause?"**

Sara Jones: "There is plenty that a regular citizen can do to support innocence work. To start, I encourage people to explore our website to learn more about wrongful convictions, their causes and consequences, and our efforts to free innocent people. To stay up to date with ongoing cases and opportunities to volunteer or engage, people can sign up for our monthly e-newsletter and follow us and share our content with others on social media (Facebook,

Twitter, Instagram, and YouTube). The public can help us spread the word to people in prison (state or federal) about our services and encourage anyone they know who is actually innocent of the crime for which they were convicted to write to us and apply for our assistance. Because of the number of requests for help we get, the strict legal standards applied in these cases, and the specificity of our mission, there are criteria we use in determining which cases we can take forward into further investigation. These are listed on our website. More generally, all people can engage with our 'calls to action' to contact legislators, support a client, or other opportunities that come up as we seek system changes and to support newly freed people. These types of opportunities can be found on our social media platforms and in our newsletter.

"Of course, donating what you can to help fuel our work makes a huge difference to us and our clients. Investigations can cost up to several thousands of dollars for things like scientific evidence testing, expert consultations, travel to interview witnesses, and attorney time. There is no charge to our clients for the work we do.

"Community members can also volunteer with us! We are often seeking volunteers who can support our events or can provide nonlegal professional services such as photography, video production, public speaking, graphic design, and administrative support. There's a volunteer interest form on our website. Finally, anyone can invite a representative from GN-IP to speak at your company, school, community organization, place of worship, etc., to discuss our work and the criminal-legal system."

Great recommendations for us all to get involved!

Amyas is poisoned by coniine in *Murder in Retrospect*. What is the science behind this poison? Coniine is an alkaloid present in poison hemlock, the yellow pitcher plant, and fool's parsley. It's toxic not only to humans but also to animals, as it causes respiratory paralysis in whoever ingests it. This poison has been used in fiction and in real life for centuries including in the death of Socrates in 399 BCE, who was sentenced to die by drinking a mixture containing poison hemlock. He was convicted of corruption of youth and impiety toward the gods.[12] Although coniine was used to carry out death sentences, it was also used to treat arthritis, cure rabies, and as an antispasmodic.[13]

The police believe Caroline was jealous of cheating and that was her motive for murder in *Murder in Retrospect*. How often is this the case? A study in 1994 found that 15 percent of people reported that they had been subjected to physical violence at the hands of a jealous partner. "Men are responsible for the majority of the killings and serious injuries resulting from jealousy, though this may reflect less a quality of male jealousy and more the qualities of male aggression."[14] Jealousy isn't always destructive and violent, though, and can be an important emotion to experience to better understand our relationships, our commitment to them, and our own feelings.

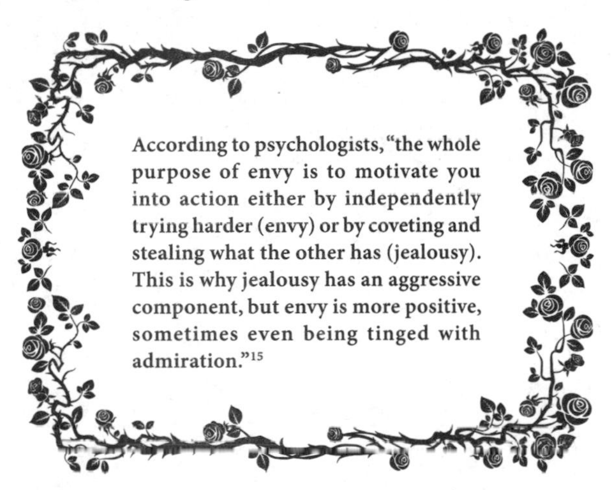

According to psychologists, "the whole purpose of envy is to motivate you into action either by independently trying harder (envy) or by coveting and stealing what the other has (jealousy). This is why jealousy has an aggressive component, but envy is more positive, sometimes even being tinged with admiration."[15]

Caroline is innocent of the crime in *Murder in Retrospect* and, thanks to Poirot's keen investigative skills, her legacy is not tainted. Thanks to the Innocence Project, this can be true in real life as well.

CHAPTER TEN

Crooked House

"There was a crooked man and he went a crooked mile,
He found a crooked sixpence against a crooked stile;
He bought a crooked cat, which caught a crooked mouse,
And they all liv'd together in a little crooked house."[1]

Crooked House (1949) isn't the first time Agatha Christie imbues her murderous plot with nursery rhymes. *Murder in Retrospect,* also known as *Five Little Pigs,* often mentions the familiar childhood classic that involves a piggy crying wee-wee-wee all the way home, and the murderer in *And Then There Were None* relies heavily on the guidance of "Ten Little Soldiers." In the first chapter of *Crooked House,* its namesake is used by Sophia Leonides to describe her family estate: "And they all lived together in a little crooked house. That's us. Not such a little house either. But definitely crooked—running to gables and half timbering!"[2] As before, Christie masterfully utilizes a harmless children's song to create dread in the reader's confrontation of the Leonides home. When the narrator, Sophia's almost-fiancé Charles, visits the grand estate after the murder of Sophia's grandfather, Aristide, he confirms this sense of crookedness:

> The curious thing was that it had a strange air of being distorted—and I thought I knew why. It was the type, really, of a cottage, it was a cottage swollen out of all proportion. It was like looking at a country cottage through a gigantic magnifying glass. The slant-wise beams, the half-timbering, the gables—it was a little crooked house that had grown like a mushroom in the night!"[3]

This physical distortion is a manifestation of the mansion's tumultuous insides, which have been converted into separate living quarters for the Leonides family. It also conveys, much like a dark, gothic castle in horror literature, the unrest of both the members of the family and the world outside of them. Due to the devastation of World War II, Roger and his wife Clemency lost their house in London and are forced into the suburbs to reside in the upstairs of the "crooked house."

Laurence Brown, the tutor employed by the family, is revealed to be an objector to the war. He skipped out on combat, and Christie portrays him as rather scared, ineffectual, and helpless. Laurence explains his torment:

> "All right then," he burst out. "What if I was—afraid? Afraid I'd make a mess of it. Afraid that when I had to pull the trigger—I mightn't be able to bring myself to do it. How can you be sure it's a Nazi you're going to kill? It might be some decent lad—some village boy—with no political leanings, just called up for his country's service. I believe war is *wrong*, do you understand? I believe it is *wrong*."[4]

While this sentiment is understandable, the way Laurence is often depicted as a trembling rabbit in a trap points to the complication of World War II's call to duty. Prewar England's "crooked" leanings are clearly on the mind of Christie's characters and, understandably, the novelist herself.

Many British women were left without husbands in the wake of the second World War, much like Agatha's own daughter, Rosalind. Yet, few were as privileged. Domestics had to find new ways to financially contribute to their household, stepping away from the class traditions and gender roles that had previously gripped them:

> Domestic service had a compelling presence in British economic, social, and cultural life. For the first half of the century, it employed the largest numbers of women of any labor market sector in Britain. Predominantly female, these servants worked in other people's homes, where they did not only do the dirty work but also formed deep attachments to those they worked for and lived out their lives under the same roof as their employers.[5]

As noted in *Agatha Christie: The Woman and Her Mysteries*, Sophia and the other women in *Crooked House* represent a new generation:

> In Sophia Leonides, Agatha Christie portrays an unusual woman who will carry the family fortune into the postwar era. It is remarkable that all the female characters in *Crooked House* except Brenda are exceptionally able and talented. Magda, a successful actress, and Clemency, an equally successful scientist, both take dominant roles in their marriages.[6]

While this is true, the depleting domestics left certain middle- and upper-class women, like Sophia's aunt, Edith de Haviland, with concerns for their future: "Don't know what we'd all do without nannies. Nearly everybody's got an old nannie. They come back and wash and iron and cook and do housework. Faithful."[7] The attention to servants continues as Charles inquires how the domestics operate in a house full of different family groupings.

In her book, *Knowing Their Place: Domestic Service in Twentieth Century Britain*, Dr. Lucy Delap "fascinate[s] . . . the contemporary reader [with] detailed analysis of the shock and loss (and this is not to overdramatize the case) that middle class families felt at the absence, or much curtailed presence, of domestic service in the post-war years, when those brought up by nanny became expected to look after their own households."[8]

This resonates in *Crooked House*, as the "nannie" of the Leonides household is the second murder victim. Like Aristide, she is killed by poison, leaving the others to scramble to find out who the culprit is before they are next. They are also left with few servants, a mirroring of the societal changes Christie would have been experiencing at the time.

As we've come to know, Christie is the master of a surprise ending, and *Crooked House* is no exception. There is no Poirot or Marple to suss out the culprit, therefore it takes a journal of confession to know that young Josephine has killed both her grandfather and her nannie! At only thirteen years old, this is quite a revelation; one that her aunt Edith decides to rectify. With the promise of ice cream, Edith takes Josephine on a deadly car ride. Edith believes that death for both herself and her murderous niece is the only answer.

We can't say Christie didn't give us clues along the way that it was Josephine who had done the devious deeds. Charles, the narrator who is an outsider of the Leonides family, discusses the case with his father, referred to as "Old Man," who happens to be a police inspector. They are ruminating on motive; whether pure hate is a worthy reason to kill someone. Charles's father makes an interesting reference: "When you say hate, I presume you mean dislike carried to excess. A jealous hate is different—that rises out of affection and frustration. Constance Kent, everybody said, was very fond of the baby brother she killed. But she wanted, one supposes, the attention and love that was bestowed on him."[9] The true story of Constance Kent culminated on a summer evening in 1860 in Devon, England. Her little brother, four-year-old Francis, was found dead in the outhouse, his throat slashed. Although it was unfathomable, sixteen-year-old Constance was the prime suspect, as she had a missing night dress, and no alibi. Because of insufficient evidence, she was let go. Five years later, repentant, Constance admitted to the murder. After her execution sentence was commuted by Queen Victoria, she served her time until age forty-one, then headed to Australia. She changed her name, volunteered as a nurse, became a nun, and died at the ripe old age of one hundred! In retrospect, this inclusion of the real murder by Constance Kent is a clever clue tucked into *Crooked House*, reminding us that teenage girls are capable of murder.

At the beginning of the film version of *Crooked House* (2017), a syringe is used to inject someone with an unknown substance. What is the history of syringes and needles? Syringes were seen in Greek and Roman literature and were described by a writer in 129 to 200 CE.[10] The first hypodermic needle in history wasn't seen until 1844. Francis Rynd used a small tube and a cutting device to allow morphine to flow out of the needle by the force of gravity and by 1854 other doctors invented the plunger syringe to quicken injections and take blood samples.[11] Modern-day disposable syringes and needles became common in the mid-1950s.

Charles feels that taking the case of Mr. Leonides's death in *Crooked House* would be unethical. According to McClain Investigations:

> Nothing speaks louder, or can have more consequences, than actions. Analyze your client, the case, subject, and your entire investigation from all of these perspectives, and the courts. It will make you both successful and responsible. The actions you take in your actions must be legal. Your actions must also be justified, including necessity and costs, and without bias to any person.[12]

It's important to always follow the law in investigations, don't put any party in jeopardy, and communicate openly and honestly in all interactions.

It is determined that Mr. Leonides's death in *Crooked House* was caused by eserine. How would this have affected his body? The drug is commonly used to treat delayed gastric emptying and glaucoma but can cause vomiting, diarrhea, and abdominal pain. In extreme cases, an overdose of the drug can cause seizures, respiratory arrest, and death.[13] Eserine, also known as physostigmine, was used to treat Alzheimer's disease for a time, but trials didn't continue due to the high rate of side effects.[14]

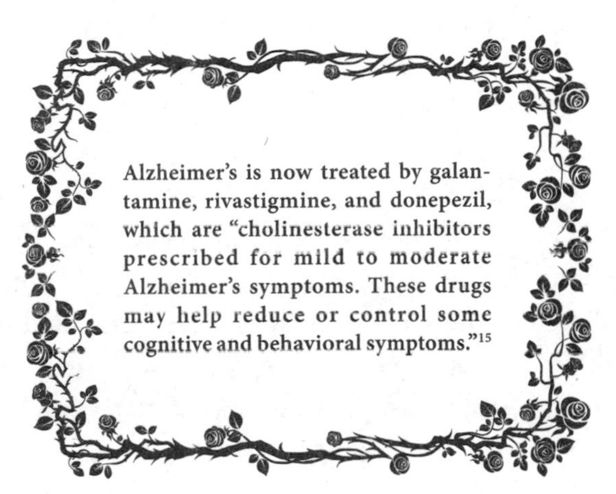

Alzheimer's is now treated by galantamine, rivastigmine, and donepezil, which are "cholinesterase inhibitors prescribed for mild to moderate Alzheimer's symptoms. These drugs may help reduce or control some cognitive and behavioral symptoms."[15]

Edith's (Glenn Close) character in the film is introduced firing a shotgun to get rid of moles. She claims that moles are hemophiliacs. Is there truth behind this? According to pest experts, moles are not hemophiliacs (and they're not blind, either!). They do have more blood than average due to their underground living environments, but clot as normal if bleeding occurs. If you have a mole problem in or near your home, it's recommended that you install a wall or fence to keep them out, reduce watering of your garden, and eliminate weeds in your garden which attract worms (80 percent of moles' food source). This is also a good time to use some hair of the dog. Not the kind you think! Putting dog hair in a molehill apparently irritates moles, and they are less likely to return.[16]

Speaking of "hair of the dog," Magda (Gillian Anderson) claims she'll get rid of her headache by using this method. In this context, it means consuming more alcohol after a night of heavy drinking. How does additional alcohol consumption help or hurt a hangover? Although this act is seen in the modern era, it's not a new concept. In fact, it first appeared in print in 1546! According to the Wake Forest Baptist Medical Center, "in medieval Europe, when astrology and blood-letting were frequently employed in the diagnosis and treatment of disease, one therapy for rabies was to place some pieces of hair from the rabid dog onto the victim's bite wound."[17] This method was proven not to work (much like using alcohol to cure a hangover).

How, historically, have children been shaped by the practice of being raised by nannies or other alternate caregivers? Nannies have been around for centuries and were linked to social class. Affluent people were able to have wet nurses or similarly named women look after their children immediately after birth. By the 1700s, the term "nanny" was coined, and the duties changed to include things like housekeeping. What does science say about children spending time with people other than their parents? Studies have shown that children ages three and up who attend daycare, or other similar settings, have improved cognitive skills, like literacy and math, as well as improved social skills[18] while those who are between zero and twelve months were found to have no boost in social skills and may have worse behavior later when attending school.[19] After much research, one scientist concluded:

> The best behavioral and cognitive outcomes come from starting half-days in daycare around [age] two and a half. Switching to full days provides no benefits and long days may worsen behavior

until around four. Before two and a half, any relative as carer gives the best outcomes. Failing that, nannies are probably better than childminders (in-home daycare) and both are certainly better than daycare centers. All of the negative effects of nonrelative childcare are more pronounced for younger children; childcare choices in the first twelve months make the most difference, as children are particularly dependent on their carers then.[20]

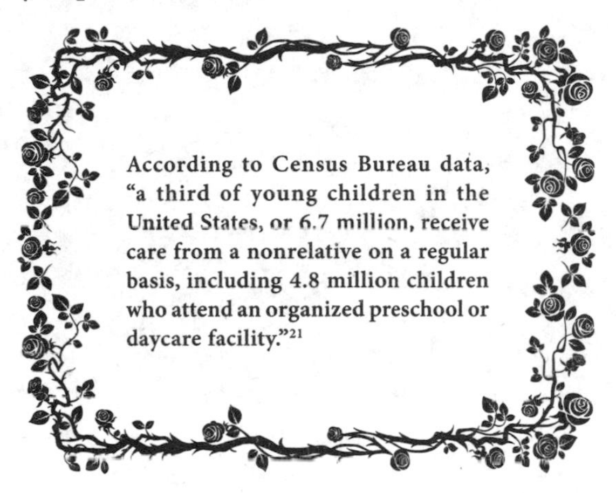

According to Census Bureau data, "a third of young children in the United States, or 6.7 million, receive care from a nonrelative on a regular basis, including 4.8 million children who attend an organized preschool or daycare facility."[21]

Charles in *Crooked House*, like other investigators, takes notes when asking questions. How does writing down information help our brains retain it? Researchers have found that writing notes by hand helps our brains preserve information more than typing the same information into a document. The physical act of handwriting helps our brains process information and allows us to understand ideas more thoroughly. Cognitive neuroscientists Fergus Cralk and Robert Lockhart discovered when people visually represent knowledge in their note-taking, like creating drawings or charts, they deepen their comprehension of concepts such as cycles and relationships.[22] This is why it's important to learn note-taking strategies from a young age. For example, I (Kelly) make sure to tell my students to write things in their own terms that they will understand instead of writing things down verbatim. It's also important to connect new information to old within your note-taking process, like putting an asterisk next to a new concept that relates to something from a previous note. Last, I encourage students to review or rewrite their notes, especially in areas they need to do further research in, so that they become clearer through the process.

"There are about 391,000 species of vascular plants currently known to science, of which about 369,000 species (or 94 percent) are flowering plants, according to a report by the Royal Botanic Gardens, Kew, in the United Kingdom."[23]

Clemency is a chemist and plant toxicologist. What do people in this profession do? The word "toxicology" comes from the Greek words for poison (*toxicon*) and scientific study (*logos*). Plant toxicologists combine the elements of many scientific disciplines to understand the harmful effects of chemicals on living organisms. It could be argued that toxicology is the oldest scientific discipline, since humans have always needed to know which plants were safe to eat and which could be poisonous. Modern toxicologists study the molecular, biochemical, and cellular processes responsible for diseases caused by exposure to chemical or physical substances, design and carry out controlled studies of specific chemicals to determine the conditions under which they can be used safely, assess the probability, or likelihood, that particular chemicals, processes, or situations present a significant risk to human health and/or the environment, and assist in the establishment of rules and regulations aimed at protecting and preserving human health and the environment.[24]

Josephine ends up being the killer in *Crooked House* because she was angry at her grandfather for making her stop taking ballet lessons. Children have been known to kill in the past, and their motivations range from the need to feel powerful to wanting to know what it feels like to kill someone. Read more about this disturbing history in our book, *The Science of Monsters* (2019).

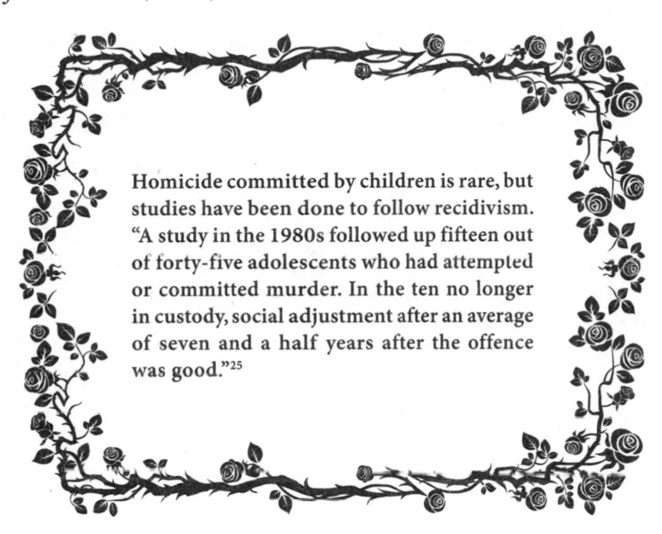

Homicide committed by children is rare, but studies have been done to follow recidivism. "A study in the 1980s followed up fifteen out of forty-five adolescents who had attempted or committed murder. In the ten no longer in custody, social adjustment after an average of seven and a half years after the offence was good."[25]

CHAPTER ELEVEN

A Murder Is Announced

In *A Murder Is Announced* (1950), the citizens of Chipping Cleghorn are intrigued by an unusual advertisement in the *North Benham News and Chipping Cleghorn Gazette*. Among the ads for farming equipment for sale and people seeking domestic help is this: "A murder is announced and will take place on Friday, October 29th at Little Paddocks, at 6:30 p.m. Friends please accept this, the only intimation."[1] This strange addition to the local paper is a surprise to those living at Little Paddocks, like mistress of the house Letitia Blacklock, as well as to her friends and neighbors.

Those who read the announcement assume it to be an elaborate way to invite them over for a "murder party" in which they will all play a game. They speculate on what sort of parlor game it might be. Will someone be chosen as detective? Will the lights be turned off as they play dead?

Parlor games, a common way to pass the time before we were all absorbed in our iPhones, had become more macabre at the beginning of the twentieth century. The "wink game" (which we both remember playing at church-youth events) involves death-by-wink in which the murderer works to conceal their identity as the bodies fall. If you were anything like us, you died most dramatically, tongue lolling! The first board game to adopt the subject of murder was the Jury Box in 1937.

> Unlike modern murder mystery party games, which are elaborate affairs that sometimes require extravagant character portrayals, The Jury Box is a more informal, easy to play party game. The players are members of a jury who get to examine evidence against an accused person. They each then try to unravel the case and decide if the defendant is guilty or innocent.[2]

Murder games were rising in popularity in this postwar era, including the still-popular board game Clue. Invented by British musician Anthony Pratt, Clue endures as one of the most recognizable and well-loved board games in the world. Bored from long nights spent during World War II air-strike blackouts, Pratt and his wife, Elva, brainstormed the characters, weapons, and setting for the game. In 1947, Pratt sold the patent to Clue (or Cluedo, as it's known in the UK) to Waddingtons, and to Parker Brothers for North American distribution. But it wasn't until the war shortages were reduced in 1949, one year prior to *A Murder Is Announced*, that Clue was introduced. "Nicolas Ricketts, curator of table games at the Strong Museum of Play in Rochester, New York, stated that Agatha Christie's bestselling mystery novels, which came out around the same time as Clue, likely boosted the game's appeal. And the game's simplicity makes it broadly appealing to adults as well as children."[3] With an obvious nod to Christie's mystery novels, Clue's characters, from Colonel Mustard to housekeeper Mrs. White, have been a staple of modern media. It was even developed into one of our all-time favorite movies starring Tim Curry and Madeline Kahn in 1985!

With the rise of murder games in the 1950s, it's no wonder that the fictional residents of Chipping Cleghorn assumed the newspaper's advertisement was a fun, inventive way to announce a game. Colonel Easterbrook believes he knows how the party will go, and as a former investigator is certain he would make a good pseudo-detective:

> It can be very good fun if it's well done. But it needs good organizing by someone who knows the ropes. You draw lots. One person is the murderer, nobody knows who. Lights out. Murderer chooses his victim. The victim has to count twenty before he screams. Then the person who is chosen to be detective takes charge. Questions everybody. Where they were, what they were doing, tries to trip the real fellow up. Yes, it's a good game—if the detective—er—knows anything about police work.[4]

Of course, the irony is that not only is it *not* a game, but it is also *not* solved by a detective. Yes, Inspector Craddock works tirelessly to find out who attempted to kill Letitia Blacklock in a staged robbery, as well as who subsequently murdered Dora Bunner and Amy Murgatroyd, but it is clever Miss Marple who notices the small clues that add up to the unlikely culprit!

Visiting Chipping Cleghorn from her village of St. Mary Mead, Miss Marple is once again left to take the investigation into her own hands. She takes to writing clues on a notepad, ordering her thoughts in a way that Hercule Poirot would stubbornly hold inside. They are curious terms, almost stream-of-consciousness, that to anyone else make little sense: "Lamp, violets, where is bottle of aspirin? Delicious Death. Making 'enquiries.' Severe affliction bravely borne. Iodine, pearls, Lotty, Berne, old-age pension."[5]

When Miss Marple goes missing for a time, her niece Bunch Harmon and Inspector Craddock try to make sense of her scribbled notes. Naturally, we as the reader work alongside them to make connections to the clues. This is one of the many reasons why Agatha Christie novels are so appealing: They are interactive! Unlike a passive experience in which the drama unfolds before us, Christie provides breadcrumbs that, if we are clever enough, will lead us, like Miss Marple, to the killer. No matter how many of her novels I (Meg) read, she has the ability to surprise me.

In *A Murder Is Announced*, Christie plays with the nature of identity. If Facebook or Instagram existed in 1950, the plot of the novel would disintegrate, as no one is able to find photos to verify Letitia Blacklock is actually her sister, Charlotte, disguising herself (and committing cold-blooded murder) for money. This paranoia, that interlopers in villages could be nefarious, deepened after World War II, as the rural ways of England were changing rapidly. Miss Marple explains:

> Fifteen years ago, one knew who everybody was . . . But it's not like that anymore. Every village and small country place is full of people who've just come and settled there without any ties to bring them. The big houses have been sold, and the cottages have been converted and changed. And people just come—and all you know about them is what they say of themselves. They've come from all over the world. People from India and Hong Kong and China, and

> people who used to live in France and Italy in little cheap places and odd islands . . . they don't wait to call until they've had a letter from a friend saying that the So-and-So's are delightful people and she's known them all their lives.[6]

In *A Murder Is Announced,* it's not just the English who are grappling with changes to their ways of life. Mitzi, the housekeeper at Little Paddocks, is a refugee from Europe. An educated young woman who has lost her entire family to the ravages of war, Mitzi is the most profoundly affected by the murder, and threat of harm, around her. She is a product of her time, a woman clearly grappling with what we now know is PTSD.

A murder is announced in the newspaper in this novel and the townsfolk think it's a cheeky way of inviting everyone over for a party. Little do they know, a murder will be committed. Have any crimes been announced beforehand throughout history? Multiple manifestos have been published online before mass killings in recent years, including a shooting in New Zealand in 2019,[7] a mass shooting at a Walmart in El Paso, also in 2019,[8] and a Buffalo supermarket shooting in 2022.[9] Infamous serial killers like the Zodiac Killer and the Son of Sam both sent or left letters that mocked law enforcement and admitted to their crimes. Read more about both serial killers in our book, *The Science of Serial Killers* (2021).

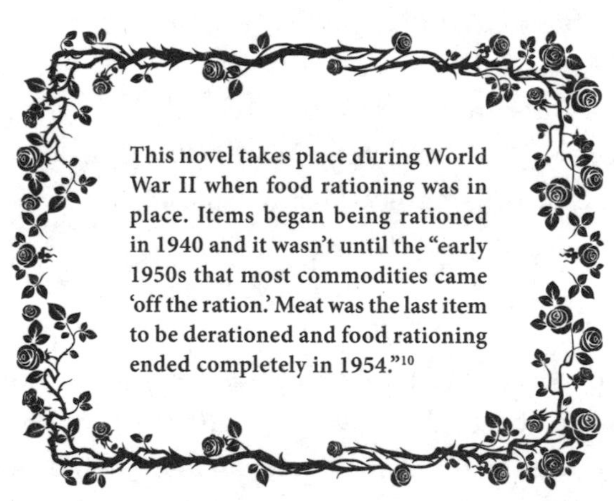

This novel takes place during World War II when food rationing was in place. Items began being rationed in 1940 and it wasn't until the "early 1950s that most commodities came 'off the ration.' Meat was the last item to be derationed and food rationing ended completely in 1954."[10]

Can we predict crimes before they happen? Perhaps. Philip K. Dick coined the term "pre-crime" in his 1956 short story, "The Minority Report," about a law enforcement agency that is able to identify and stop people who will commit crimes in the future. Though this story was science fiction, science has used techniques to try to recognize potential criminals before they commit crimes since the late nineteenth century. Cesare Lombroso, a criminologist and physician, believed that there are "born criminals" who can be recognized on the basis of physical characteristics including a sloping forehead, asymmetry of the face, and excessive length of the arms, to name a few. Although many of his theories were disproven, he left a legacy in the world of criminology and psychiatry. Today, technology and software allow law enforcement agencies to track patterns and analyze data that may predict criminal behavior. It's not a perfect system yet, though. Experts cite worries about privacy, stereotyping, and racial profiling when trying to predict who will commit a crime.[11]

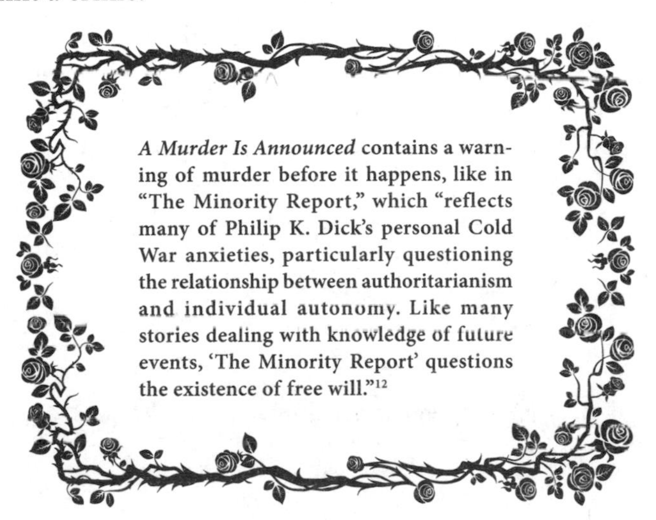

A Murder Is Announced contains a warning of murder before it happens, like in "The Minority Report," which "reflects many of Philip K. Dick's personal Cold War anxieties, particularly questioning the relationship between authoritarianism and individual autonomy. Like many stories dealing with knowledge of future events, 'The Minority Report' questions the existence of free will."[12]

Disguises are brought up in *A Murder Is Announced* and how people use them to deceive others about their appearance and identity. People often joke about how superheroes in movies can put on a simple pair of glasses or a baseball cap and they are unrecognizable to their friends and acquaintances. Is it possible to deceive others with a simple disguise? Jonna Mendez, the Central Intelligence Agency's former disguise chief, worked for years with spies who needed to disguise themselves. She focused on the most obvious features of the person and changed them, like straight hair to curly or making a young person look older, as well as on how they walk and carry themselves. "You want to be the person that gets on the elevator, and then gets off, and nobody really remembers that you were even there. That is a design goal at the disguise labs at CIA."[13]

We may not recognize someone if they have on an excellent disguise but what if we don't recognize them at all, even though we know them? This phenomenon is known as prosopagnosia, or face blindness. This condition gained media attention in 2022 when actor Brad Pitt revealed that he suffers from it. Face blindness can occur from birth or can develop after suffering a head injury or stroke. Sufferers are unable to recognize people, even their own family members, and may have difficulty forming relationships. Another actor, Shenaz Treasury, wrote, "I've always felt so ashamed that I mix up people and can't recognize faces of people—even close friends if I see after a few years—I can't recognize them. This is a real brain issue. Please be kind and understand."[14] What are some tips for helping you remember a person that you just met? Make sure to repeat their name immediately after they share it and say it at least two more times in conversation. Link something about that person's appearance to their name so you'll have an easier time remembering it. Finally, go back over your conversation with that person and repeat their name to yourself to help lock it into your memory.

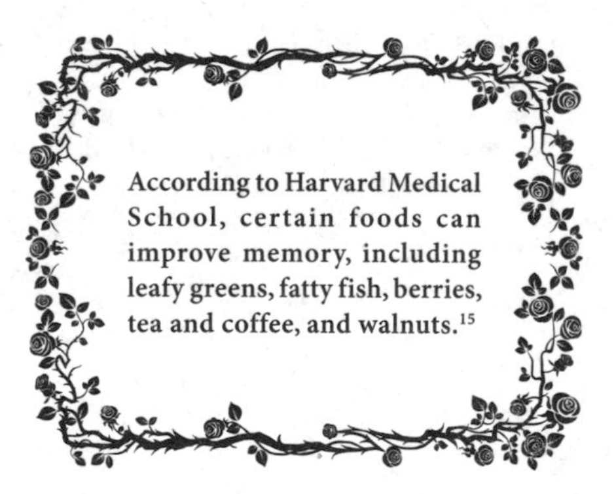

According to Harvard Medical School, certain foods can improve memory, including leafy greens, fatty fish, berries, tea and coffee, and walnuts.[15]

Soothing colors brought to the house in *A Murder Is Announced* are said to calm nerves after the crime. How do colors play a role in our moods and emotions? Chromotherapy was first discussed in 1025 CE by Avicenna, a Persian physician, who believed color to be an observable symptom of disease. Although color therapy is often regarded as quackery, some holistic healers today promote this practice, "the science of using colors to adjust body vibrations to frequencies that are said to result in health and harmony."[16] The presence of color has been proven to affect mood, but psychologists believe it is based on our own personal preferences, experiences, and culture. Even so, it doesn't stop artists, designers, and marketing experts from using color to try to evoke an emotional response from us!

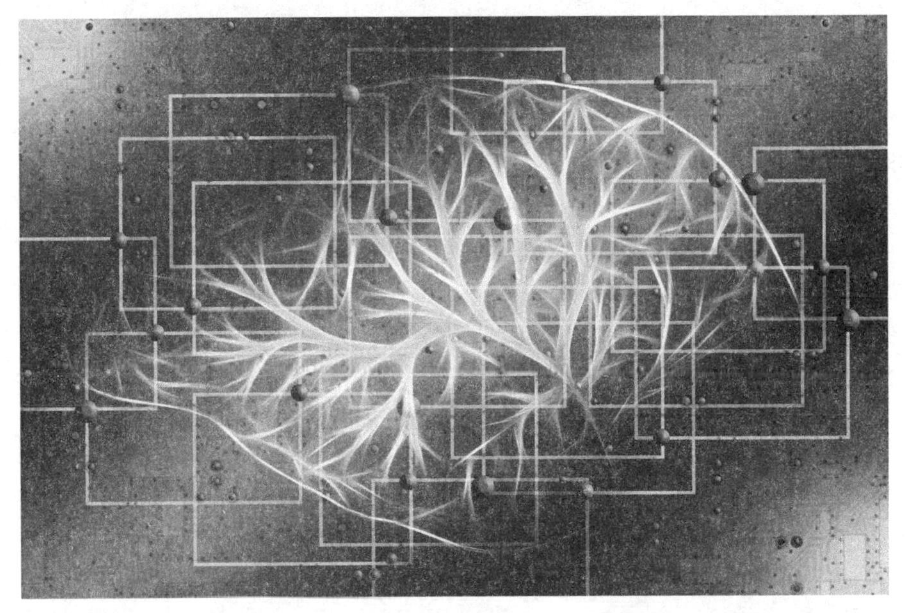

"According to scientists, memories are formed as a result of connections between neurons in the brain. New connections (synapses) are formed each time a new activity is learned. The more a person participates in a particular activity, the stronger the synapses and associated memories tied to the activity become."[17]

A witness is interrogated in *A Murder Is Announced*. How can questioning influence memory? According to Dr. Brian Fitch, a lieutenant with the Los Angeles county Sheriff's Department:

> Interviewers should avoid leading or suggestive questions whenever possible. They should use open-ended questions that require detailed responses and that do not offer built-in "default" answers.

> Closed-ended questions restrict the amount of information provided, whereas open-ended questions offer no assumptions and place no restrictions on answers. Investigators may use follow-up and closed-ended questions after an interviewee has provided as many details as possible through open-ended questioning.[18]

Poison is used to kill the next victim in Christie's story. What poisons could possibly work within seconds as this one did? Cyanide, whether it's ingested or inhaled, can cause death within minutes while ricin, made popular through the television series *Breaking Bad* (2008–2013), takes anywhere from thirty-six to seventy-two hours to kill. Sarin gas is lethal within five to ten minutes, and anthrax kills the infected person three to five days after inhalation. Another quick poison? Tetrodotoxin. "Found in the organs of puffer fish (the famous Japanese delicacy, fugu), tetrodotoxin persists even after the fish is cooked. If the toxin is consumed, paralysis and death can strike within six hours. Up to five Japanese die from badly prepared fugu every year."[19] We plan to avoid that delicacy to be on the safe side!

Miss Marple pours water on a lamp's electrical cord in *A Murder Is Announced*, and the power goes out in the entire house. Is this accurate? If a cord is in good shape, spilling a bit of water on it shouldn't be a problem. But if the wire is exposed, it could cause a short circuit which could blow a fuse (as in Miss Marple's case) or it could even start a fire. This is because the presence of moisture increases the current in an electric circuit quickly. If you do get wiring significantly wet, experts recommend turning off the electricity and calling an electrician to assess the damage.[20] It's better to be safe than risk even more damage by having wiring collect more moisture which will cause corrosion and rust.

A goiter removal is mentioned in *A Murder Is Announced*, and it becomes a major plot point. Why would someone need this procedure done? A goiter is a noncancerous enlargement of the thyroid and may need to be removed if is uncomfortable, causes difficulty breathing or swallowing, or causes hypothyroidism. According to the Mayo Clinic, the surgery is typically performed under general anesthesia and carries minimal risks including a hoarse voice, some bleeding, and neck pain.[21]

Charlotte assumes the identity of her dead sibling, Letitia, in this novel. Has this ever happened in history? It has! A 2020 study found that "the death of a close relative has been linked to a variety of adverse health consequences among bereaved family members. An excess risk of morbidity and mortality was observed among individuals who experienced the loss of a spouse, parent, or child," and with the loss of a twin, specifically, it greatly increased the risk of psychiatric disorders.[22] Researchers concluded that this risk was not only based on genetic factors of twins but also on their shared early life experiences and attachments. A real-life case of a sibling assuming her sister's identity for money occurred in Oakland, California. In 1997, Sarah Allman was arrested for impersonating her dead sister, who she murdered, in order to collect her pension checks. Police were unsure how long her sister, Stevie, was dead but it could have been up to two years.[23]

After reading this story, we've been inspired to hold a murder mystery party! But please, hold the murder.

CHAPTER TWELVE

They Do It with Mirrors

At the start of *They Do It with Mirrors* (1952), Miss Marple meets with her longtime friend, Ruth Van Rydock. The reader is immediately met with Ruth's vivid physical description.

> She was exquisitely corseted. Her still shapely legs were encased in fine nylon stockings. Her face, beneath a layer of cosmetics and constantly toned up by massage, appeared almost girlish at a slight distance. Her hair was less grey than tending to hydrangea blue and was perfectly set. It was practically impossible when looking at Mrs. Van Rydock to imagine what she would be like in a natural state.[1]

As both Ruth and Jane Marple gaze into a mirror, Ruth asks, "Do you think most people would guess, Jane, that you and I are practically the same age?"[2] While we would roll our eyes at such pomposity, kind and white-haired Miss Marple reassures her friend that no one would ever guess!

This struck a chord, because in the last book we researched, *A Murder Is Announced*, there were quite a few mentions of aging. For instance, "All old women look alike."[3] Therefore, elderly Miss Blacklock was able to pretend to be her sister under no suspicion. This notion seems to be in stark disagreement to the physical differences of Ruth and Miss Marple, yet Ruth is quick to point out, "They know I'm an old hag all right! And, my God, do I feel like one!"[4] These meditations on aging, from the mouths of fictional characters, made us wonder if Christie was describing her own frustrations about becoming an older woman.

Agatha Christie turned sixty-two years old the year *They Do It with Mirrors* was published. She was a grandmother and had established her throne as the Queen of Crime. Rich, successful, and well-loved by fans and family, Christie had what everyone might hope for. But she had been raised in an affluent society that valued physical looks, especially women's:

> In her old age, Agatha Christie looked back with understandable nostalgia to the period in her late teens and early twenties when she was slim and active, with thick, wavy, waist-long blonde hair and the delicate skin and sloping shoulders that were all the rage. She never ceased to mourn this youthful self: in later life she was to make constant rueful comments about her weight, and had a horror of being photographed.[5]

The Queen of Crime was undoubtedly human and, like Miss Marple, was perhaps comparing herself to other women. Because of her prolific nature, we can read alongside Agatha Christie as she ages, mining clues in her fiction of how she might have felt.

Like many of her novels, *They Do It with Mirrors*, known in the UK as *Murder with Mirrors*, has been adapted to screen, most notably in 1985 for British television, starring none other than American film legend Bette Davis. Just a few years before she died, Davis portrayed Carrie Louise Serrocold, another longtime friend of Miss Marple who runs a home for delinquent boys on her mansion's property.

During World War II, many mansions in rural England were requisitioned by the government to use as maternity wards, military hospitals, and even to store weapons and supplies. One example is the mansion known as Shardeloes in Amersham, Buckinghamshire:

> Home to the descendants of Elizabethan explorer Sir Francis Drake, it was requisitioned for medical purposes. It had been repurposed to become a maternity home for evacuated mothers. A local paper reported that "it was offered to the Ministry of Health as a maternity hospital some time ago, and on the outbreak of war it was converted within twelve hours—the furniture stored in two of the rooms, the pictures removed and the wall spaces labeled, the library boarded up and provision made for fifty beds."[6]

Postwar, it wasn't unheard of for the owners of these grand estates to continue aiding those less fortunate. Carrie Louise and her husband, Lewis Serrocold, devote their life to the cause of young men who need guidance after criminal acts. The Serrocolds live and work among these men—all suspects when Carrie Louise's stepson, Christian Gulbrandsen, is found murdered!

> [The] British juvenile justice system underwent significant philosophical changes in the first half of the twentieth century. Although there were many who clung to older ideas about the benefits of corporal punishment, the view that children and young people who broke the law should be reclaimed and rehabilitated had become the orthodox view by the passing of the 1948 Children Act.[7]

The Children Act was a rather groundbreaking law in the UK, altering how orphans, even those older than eighteen, were taken care of by the state. Committees and officers were assigned to make certain that youth, especially those displaced from war and poverty, were given the chance to thrive. It is said that this Act was set into motion after the 1945 brutal death of twelve-year-old Dennis O'Neill at the hands of his foster parents. Christie had been aware of the case, so much so that it was an inspiration for her short story, "Three Blind Mice" (1950).

The troubled boys at the fictional Stonygates Estate seem like perfect suspects—but Agatha Christie always avoids the obvious. Instead, it becomes quite clear as subsequent murders occur that it is someone within the Serrocold family who is the culprit. The title, *They Do It with Mirrors*, is an allusion to a slang term that refers to illusionists or stage musicians who divert attention to trick the audience. Finally, Miss Marple uses her wisdom and cleverness to figure out these diversions, realizing it is Lewis Serrocold who is the murderer. And money is the motive. Not only that, but Edgar Lawson, Lewis's assistant, who had been pretending to be afflicted by schizophrenia, was his son . . . and partner in murder!

Edgar attempts to make a getaway in a boat, and when it sinks, his father is given a chance to save him. It is Lewis's final act, as father and son drown together in the marsh. This dichotomy, of murderer and attempted savior, is not lost on Carrie Louise: "I'm glad it ended that way . . . with his life given in the hope of saving the boy . . . People who can be very good can be very bad, too."[8]

The television adaptation of *They Do It with Mirrors* on *Agatha Christie's Marple* (2010) begins with a home invasion and fire. The history of fighting fires in Britain and the United States has evolved over the centuries. The first organized group of firefighters originated in 43 CE, during the Roman invasion. Buckets of water were passed from one person to the next to try to slow and put out the spread of fire. During the Middle Ages, many towns in England burned down completely due to the lack of firefighters. In 1666, the Great Fire of London brought public attention to the need for standardized firefighting. The fire began in a bakery but ended up leaving tens of thousands of people homeless. The first fire insurance company was established after this incident and firefighters began responding to fires at addresses that were insured. In America, the government didn't run fire departments prior to the Civil War. By the 1800s, Americans developed the tradition of fighting fires as a grassroots collective bucket brigade. This evolved to manually operated pumps that were pulled by horses or groups of people and eventually to steam-powered pumps.[9]

An example of a fire truck from between 1870 and 1910.

What are fire statistics for the current era? According to the Center for Disease Control, the US Fire Administration, and the National Fire Protection Association (NFPA), there are an average of 358,300 home-based fires every year and the top three causes are cooking, heating equipment, and electrical malfunction.[10] Due to the change in materials used in household items like carpets, drapes, and other furniture since the 1950s, the average time to escape a house fire is now three to four minutes compared to thirty minutes in the past.[11]

The mother in *They Do It with Mirrors* takes a dip in the pond to help soothe her arthritis. What is the science behind this? To relieve the symptoms of arthritis, doctors recommend a combination of strength training, range-of-motion exercises, and endurance exercises, which can all take place in water to reduce stress on the joints.[12] Cold therapy, like soaking joints in an ice bath or taking a swim in a cool lake, can help ease discomfort from conditions like arthritis because cold naturally constricts blood vessels, reduces swelling, and numbs nerve endings associated with pain. This is why athletes are often seen taking ice baths after participating in rigorous sports. But does it work? According to a study published in *The Journal of Physiology*, "cold water immersion is no more effective than active recovery for reducing inflammation or cellular stress in muscle after a bout of resistance exercise."[13]

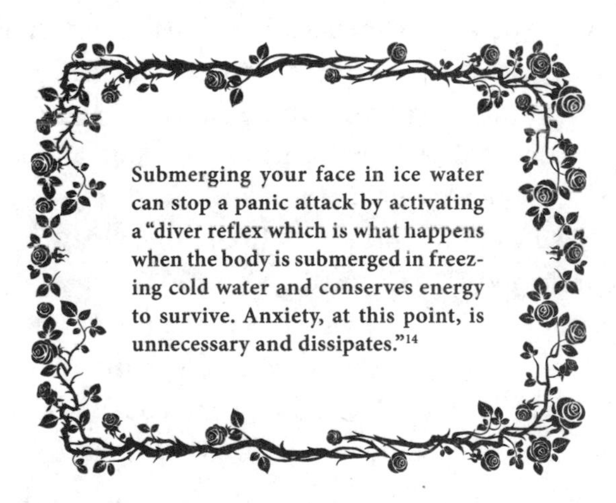

Submerging your face in ice water can stop a panic attack by activating a "diver reflex which is what happens when the body is submerged in freezing cold water and conserves energy to survive. Anxiety, at this point, is unnecessary and dissipates."[14]

Edgar is employed as a personal secretary in *They Do It with Mirrors*. He seems to suffer from some sort of delusional disorder, believing Winston Churchill to be his father. What are the medical diagnoses for cases like this? First, there is an experience that more people are encountering with the prominence of media called parasocial interaction. We start to feel like we know celebrities or fictional characters because we spend so much time watching them, listening to them, or reading about them. Meg and I (Kelly) have experienced this with people who have listened to our podcast, *Horror Rewind*, for years and interact with us as if we're old friends. This phenomenon was first described by sociologists Horton and Wohl in their 1956 study published in *Psychiatry* entitled "Mass Communication and Para-Social Interaction: Observations on Intimacy at a Distance." They concluded that "for most people, parasocial interactions with persona complement their current social interactions, while also suggesting that there are some individuals who exhibit extreme parasociality, or they substitute parasocial interactions for actual social interactions."[15]

There are both positive and negative aspects of parasocial relationships. For example, empathizing and identifying with a character or person may help us learn more about ourselves through the process. We may discover someone who has similar attributes to our own and feel "seen" in the world because of them. Other times, we may hold unrealistic expectations for a person or character and grieve when we discover that they are not everything we thought they would be. For example, someone may idolize a character for their personality traits on a television show but then find out in an interview that the actor is nothing like the character they are portraying. It's important to differentiate between what is real and what is portrayed through media.

These relationships can become a problem when they cross over into a condition known as celebrity worship syndrome. This syndrome can encompass everything from stalking someone in the public eye to committing a crime on their behalf. Treatment for this condition always begins with a mental health assessment, as this condition is often linked with depression and anxiety.[16] Transference is the term used to describe Edgar's condition and treatment, which was first used by Sigmund Freud.

This phenomenon occurs when the feelings a person has about one thing are unconsciously redirected or transferred to the present situation.[17] This often occurs in therapy settings but can occur outside of them as well.

When the investigators are questioning witnesses in *They Do It with Mirrors*, their stories differ depending on what they were able to see in the dark. How does human night vision compare to other animals? Humans' eyes naturally adjust to dark settings by opening the pupils to gather as much light as possible. Rods and cones in our retinas then send impulses to the optic nerve at the back of our brains. Rods are helpful in low-light settings but don't process color as well, so our night vision tends to be black and white in appearance. Compared to other species, our night vision is poor. Because we are primarily awake during the daytime, it takes five to eight minutes to begin to adjust our eyes to darkness and a whopping forty minutes before we reach maximum adjustment![18] Animals, on the other hand, tend to have bigger eyes and retinas as well as an extra layer called a tapetum. This layer gives animals a second chance to reflect the image back and makes their eyes glow in the dark. Nocturnal animals have more rods and cones than humans and this also helps them process images better than we can.[19]

It's believed that the character Carrie Louise Serrocold is poisoned with arsenic in *They Do It with Mirrors*. How does this poison affect the body? Symptoms of arsenic poisoning can include vomiting, diarrhea, abdominal pain, numbness, and can lead to heart disease and cancer.[20] Although the poisoning in this story has nefarious intentions, most arsenic poisoning inadvertently occurs by consuming contaminated drinking water. Arsenic is naturally found in water and regulations are in place throughout the world to monitor levels so that they are safe for human consumption. People are typically diagnosed with arsenic poisoning through urine or blood tests or by measuring the amount of arsenic in someone's hair or fingernails.

A man gets sick after eating room-temperature oysters, but Miss Marple points out that since he grew up around them, he would recognize if they were "off," concluding he must have been poisoned. How common is unintentional food poisoning? According to the CDC, "every year, an estimated one in six Americans (or 48 million people) get sick, 128,000 are hospitalized, and 3,000 die from foodborne diseases."[21] Ways to prevent food poisoning include keeping your hands and cooking surfaces clean, eating cooked foods at the proper internal temperature, keeping your refrigerator at forty degrees Fahrenheit or below, and separating raw meats from other foods.

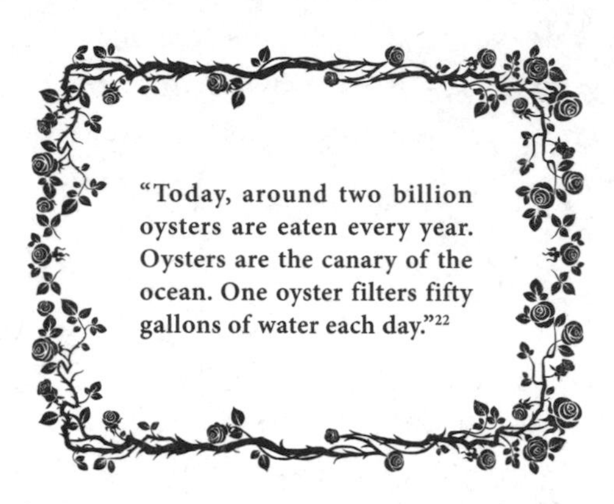

"Today, around two billion oysters are eaten every year. Oysters are the canary of the ocean. One oyster filters fifty gallons of water each day."[22]

In *They Do It with Mirrors*, the investigators time and track distances to see how long it would take someone to enter the room where a victim was murdered. How are gait and stride used in crime investigation? Forensic gait analysis is used to determine how an individual walks in a specific pattern. This can determine if the suspect was walking or running, can hypothesize their height and weight, and can be used to track against eyewitness testimony about someone's "style" of walking or compare with video surveillance.[23] Certain factors may affect a person's gait including the type of footwear they have on, if they are fatigued, if they have a medical condition or injury, their age, and their gender. All of these contribute to an investigation of a crime scene.

A chalk outline is used in *They Do It with Mirrors* to mark where the body was found after the theatre accident. Are chalk outlines still used? What is their purpose? Although a popular trope in crime procedurals on television and in film, chalk outlines are rarely used in actual criminal investigation. While they may have been used in the past, they were typically drawn for photographers to illustrate the crime scene without the body being present. The practice can potentially contaminate the area by disrupting evidence.[24]

Miss Marple refers to misdirection like the expression used to describe magic: "They do it with mirrors." Carrie Louise was never actually being poisoned; it was a misdirection.

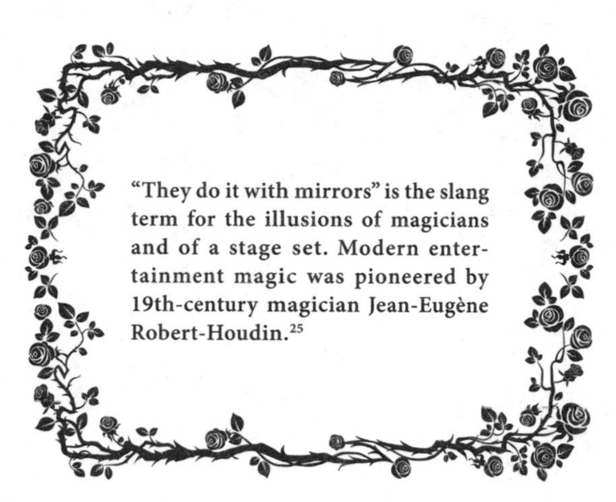

"They do it with mirrors" is the slang term for the illusions of magicians and of a stage set. Modern entertainment magic was pioneered by 19th-century magician Jean-Eugène Robert-Houdin.[25]

CHAPTER THIRTEEN
The Mousetrap

One doesn't have to look far to truly appreciate the longevity and cultural impact of Agatha Christie's play, *The Mousetrap.* The longest-running play in the world was first performed at London's Ambassadors Theatre in November 1952. *The Mousetrap* was an instant success, remaining at the 450-seat Ambassadors Theatre for over two decades (and nine thousand performances)! The show moved to St. Martin's Theatre in 1974 where it remains today, only pausing for the COVID-19 pandemic before its triumphant return in 2021.

In November 2012:

> The 25,000th performance was marked with a one-off star-studded performance, introduced by Christie's grandson Mathew Prichard and featuring Patrick Stewart, Julie Walters, and Miranda Hart. The performance accompanied the unveiling of the Agatha Christie memorial statue in Leicester Square which commemorated her great works and her contributions to the theatre.[1]

The bronze statue at the intersection of Cranbourn Street and Great Newport Street is in the shape of a large book with Christie's profile adorning the cover. On the back there are several etchings in the brass, including likenesses of Hercule Poirot and Miss Marple, as well as Egyptian pyramids and the *Orient Express*. Torquay, Christie's birthplace, is also home to a statue honoring the Queen of Crime. The bust of Christie sits on the Agatha Christie Mile, which runs along Torquay's seafront. A new piece of public art has been commissioned by the Torquay city council to be unveiled in 2023 by artist Elizabeth Hadley. Hadley chose to make a bench overlooking the sea, with a figure of Agatha Christie and her beloved dog, Peter. The design was voted on by locals and will be a permanent fixture on the Agatha Christie Mile.

While *The Mousetrap* was returning to London's West End in 2022, the film *See How They Run*, starring Sam Rockwell, debuted in movie theaters. It takes place amid the backstage goings-on of *The Mousetrap* when the director hired to adapt the play to film (Adrien Brody) is inevitably murdered. In Christie-like fashion, the quirky detectives (Rockwell, along with Saoirse Ronan) must find whodunit. "With a sprightly wit and an all-star cast to bring it to life, the movie manages to be a loving parody of theater gossips, postwar London, and Christie's murder mysteries all at once."[2] In reality, no English-language film has been made of *The Mousetrap* despite its theatrical appeal, though you can watch *Chupi Chupi Aashey* (1960), a Bengali film adaptation, or *Myshelovka* (1990) in Russian.

The Mousetrap is not just confined to the United Kingdom. It is still Canada's longest running play after a twenty-six-year run, and has been performed everywhere, even here where I (Meg) live in Southern Minnesota! A few years ago, I attended a dinner theater with my mom, and it was our first time seeing it. As Christie lovers and self-professed anglophiles, we were in heaven! And the twist worked, wowing us nearly seventy years after its initial run. Since then, my mom was able to see the official show at St. Martin's Theatre. I've added it to my bucket list as something I must do when I'm lucky enough to be in London next! And if I go, I'll be able to see an original prop that has lasted from the first performance in 1952: the clock on the fireplace mantle (the only piece to have survived seventy years and several stage makeovers)!

The Mousetrap was based on another of Christie's short stories, "Three Blind Mice," originally written in audio format for radio. The short, thrilling radio play was heard in 1947. Unfortunately, no original recording of "Three Blind Mice" exists, as it has been lost to time. Fascinatingly, the audio script was written at the behest of the Queen Mother Mary of Teck, who asked specifically for Agatha Christie to write something for the radio celebration of Her Majesty's eightieth birthday. A year later, "Three Blind Mice" was adapted for a half-hour episode on BBC, and *then* Christie altered it into a short story that was published first in *Cosmopolitan* magazine in 1948, and then in her collection *Three Blind Mice and Other Stories* (1950). After such a lengthy journey, "Three Blind Mice" would be changed once more, into a full-length play that still resonates today. In her biography, Agatha Christie discusses its peculiar genesis:

> The more I thought of "Three Blind Mice," the more I felt that it might expand from a radio play lasting twenty minutes to a three-act thriller. It wanted a couple of extra characters, a fuller background and plot, and a slow working up to the climax. I think one of the advantages *The Mousetrap*, as the stage version of "Three Blind Mice" was called, has had over other plays is the fact that it was really written from précis, so that it had to be the bare bones of the skeleton coated with flesh.[3]

Said perfectly by the Queen of Crime!

Christie goes on to explain how the name change came about:

> For the title, I must give my full thanks to my son-in-law, Anthony Hicks. When the original title could not be used—there was already a play of that name—we all exhausted ourselves in thinking of titles. Anthony came up with *The Mousetrap*. It was adopted. He ought to have shared in the royalties, I think, but then we never dreamed that this particular play was going to make theatrical history.[4]

It's believed that Anthony Hicks's suggestion of *The Mousetrap* was derived from William Shakespeare's *Hamlet*. In act III scene II, Claudius asks Hamlet the name of the play within the play (very meta!):

> **Claudius:** What do you call the play?
> **Hamlet:** The Mouse-trap. Marry, how? Tropically. This play is the image of a murder done in Vienna.[5]

As *Hamlet* is often considered the first crime thriller on the stage, *The Mousetrap* is a fitting title for such an impactful play.

The murders committed in *The Mousetrap* are motivated by a need for revenge. How often is this a motivating factor in crimes? According to a 2018 study:

> Retaliation against a partner for emotional harm has been identified as the most common reason for intimate partner violence. Retaliatory attitudes among assault-injured youth have been shown to fuel cycles of violence, and a systematic, multi-country review found that perpetrator desire for revenge was a primary motivation behind almost 40 percent of school shootings. Revenge has also been observed to be a primary motivation behind acts of terrorism.[6]

This study also cites research that shows a possible link between the need for revenge and strong cravings that found "the same reward-processing centers of the brain that activate for and are implicated in narcotics addiction also activate when people are meting out altruistic punishments, that is, willingly incurring a cost to punish others who commit perceived injustices or norm violations."[7]

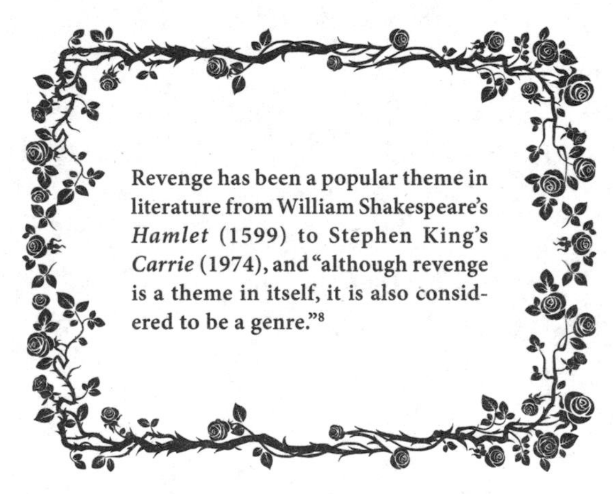

Revenge has been a popular theme in literature from William Shakespeare's *Hamlet* (1599) to Stephen King's *Carrie* (1974), and "although revenge is a theme in itself, it is also considered to be a genre."[8]

With *The Mousetrap* most famously being the longest-running show in history, we wanted to learn more about how Agatha Christie contributed to the lives of people in the world of theatre. We spoke with director Merry Renn Vaughan about her life on and off the stage.

Kelly: **"Tell us about your background in theatre!"**

Merry Renn Vaughan: "I have loved theatre for as long as I can remember. In elementary school, I was that kid who did a 'skit' for every single book report project. I just loved bringing characters to life. I did a lot of acting in high school, but my senior year, I was not cast in the final show. There was a long set of callbacks, and the director ultimately chose a blonde sophomore because he thought she fit the look of the role better. I quit theatre when that happened. The next day, I went in to speak to the director. I told him I had to be involved in the final show of my high school career and asked if I could help backstage in some way. He already had an assistant director but created the role of 'director's assistant' for me. I was even allowed to direct a couple of the scenes. It was in doing that that I discovered my true love in theatre was in directing. So, out of what I thought was one of the worst moments for me in theatre, I found my true calling.

"I left that fall for college knowing I wanted to major in directing. I completed my BFA in only three years and then jumped into the master's program—completing my first year in grad school in what would have been my final year as an undergrad. I did a quantitative audience analysis of the audience demographics at the Manion Theatre at the University of Wisconsin-Superior. I directed locally in the Twin Ports and then was asked to guest direct at the College of St. Scholastica. After that first production, I taught at Scholastica and UW-Superior and directed productions on both campuses until I became a full-time faculty member at Scholastica.

"That's the 'story' of how I got to where I am, but if you just want the facts: I have a bachelor of fine arts in theatre in directing and producing from the University of Wisconsin-Superior, a master of arts in theatre arts, also from UW-Superior, and a doctorate in philosophy in interdisciplinary studies from the Union Institute and University in Cincinnati, Ohio. My dissertation, on the ethics of Dr. Seuss, is titled *There's More to Seuss than Meets the Eye: The Social and Political Vision of an American Icon*."

Meg: **"I would love to read that!"**

Meg: **"What was the first Agatha Christie play you were involved with?"**

Merry Renn Vaughan: "The first Agatha Christie play I ever directed was *And Then There Were None*, in 2011 at St. Scholastica. So, technically that is the first Christie play I was ever 'involved' with, but I've known her work my entire life. My grandmother loved Agatha Christie. She owned every single one of her novels and had all the movies on VHS. She and my grandfather faithfully joined Hercule Poirot on PBS every Sunday night as he solved another mystery. Dame Christie was so much a part of my early life; I honestly don't remember becoming aware of her existence; she was just always there.

"However, I do recall my first encounter with Christie onstage. I was in middle school and saw a production of *Ten Little Indians* at the Duluth Playhouse. That production was magical for me; I swear the little statuettes disappeared from the mantle right before my eyes. It was because of that moment for me as an audience member that the cast of *And Then There Were None* worked really hard to choreograph the scenes so that the statues of the soldiers on the mantle really did disappear. We worked on distraction and sleight of hand as diligently as any of the other stage directions to create that same magic I had experienced. Interestingly, the 'magic' of that production was really all in my overly imaginative brain. I've spoken to cast members since that Playhouse production and they all said it wasn't real, that the statuettes didn't disappear onstage, they were just removed during scene changes.

The Mousetrap is the longest-running show in London's West End at St. Martin's Theatre.

Next season I'm directing *The Mousetrap* for CSS Theatre and I am really looking forward to building the suspense with Dame Christie again."

Kelly: **"What is your process going into directing a play? What research and preparation are involved?"**

Merry Renn Vaughan: "The first step, of course, is to choose a script. There are many things that go into this decision, but that has to be the first step. After I've chosen a script, I read it multiple times, looking at the script through a different lens each time. The first time is just to get the story. Subsequent readings are for character, set design, light design, costume design, props, etc. I am a director that doesn't really like to read stage directions because in a published script the stage directions are usually what was written into the stage manager's prompt book and they are specific to that production. But sometimes there are stage directions that are important to telling the story or that include important stage business, so I always have to make myself sit down at least once and read the stage directions all the way through.

"There is of course research that needs to be done at some level for every production. Sometimes it is really simple, like learning an obscure word or a location that you don't know about. But with historical works, there needs to be a lot of research. As an American, doing a script by Agatha Christie requires research for a number of reasons. First of all, the scripts are British and while people who live in the US and people in the UK do both speak English, there are many, many terms and phrases that not only mean different things but that are pronounced differently. Additionally, there is the historical aspect of the Christie era. It's definitely not 2022, and in order to help actors understand these things and convey them to an audience, a director has to be solid in their own understanding.

"While it is not a requirement of course, most productions of an Agatha Christie script do require actors to speak with a British accent. Learning any accent can be tricky and that is why there are dialect coaches who teach people accents for a living. I have often brought experts in to help with our productions. One of the things I do as a director to help actors who need to speak in an accent is to require those accents at all times while actors are in the building. Anyone who is working on the show needs to speak in the accent from the moment they enter the building. That way it just becomes second nature for everyone to do so. That includes me and the rest of the creative team as well as the actors."

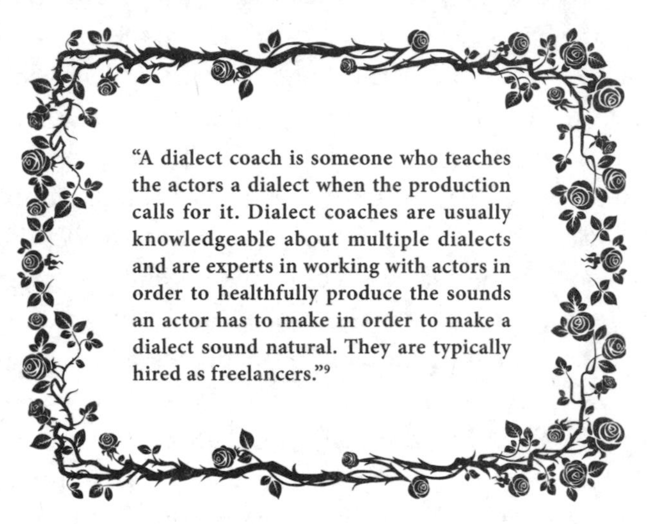

"A dialect coach is someone who teaches the actors a dialect when the production calls for it. Dialect coaches are usually knowledgeable about multiple dialects and are experts in working with actors in order to healthfully produce the sounds an actor has to make in order to make a dialect sound natural. They are typically hired as freelancers."[9]

Kelly: **"Why do you think Agatha Christie's mysteries resonate with audiences still in this day and age?"**

Merry Renn Vaughan: "I think first and foremost that Christie is a great storyteller, and everyone loves a great story. Not only does she tell a good story, but she creates great characters; the time-honored 'whodunit' that brings an air of campy fun as an audience goes along for the ride. With Christie, there's always going to be bodies, there will be a red herring or two, and someone will definitely get poisoned! There's usually a twist at the end, too. That's one of my favorite parts: can you tell her story well enough to get a gasp or two from the audience?"

Meg: **"What is your favorite Agatha Christie book/movie/play and why?"**

Merry Renn Vaughan: "I think my favorite would have to be *And Then There Were None* (or *Ten Little Indians* or *Ten Little Soldiers* depending on the era in which you experienced the show). It has all the classic Christie moves and it was my first live theatre Christie. I also like it because, like *The Mousetrap,* it stands alone in the Christie canon because it doesn't include her famous detectives. Don't get me wrong, I love Hercule Poirot and Miss Jane Marple, too, but they have their place in the world of PBS and the BBC and Kenneth Branagh has done a marvelous job recreating Poirot on the big screen. So, while the detectives are probably Christie's best-known creations, I love that *And Then There Were None* is just a good old-fashioned whodunit and lets the audience play detective to try to figure out who the killer is on Soldier Island."

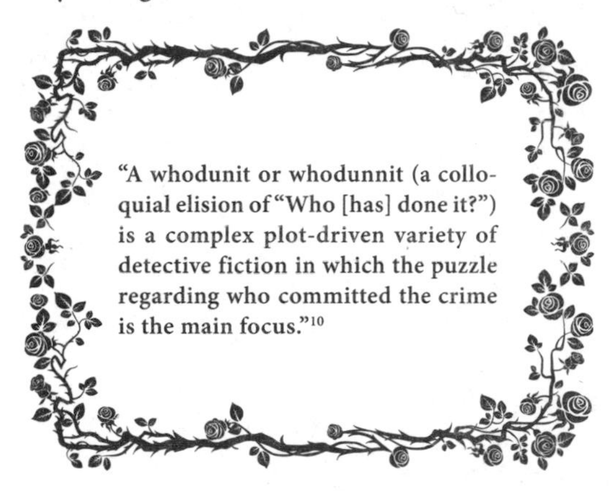

"A whodunit or whodunnit (a colloquial elision of "Who [has] done it?") is a complex plot-driven variety of detective fiction in which the puzzle regarding who committed the crime is the main focus."[10]

We had the opportunity to see Merry Renn Vaughan's production of *The Mousetrap* in November 2022, and it truly was a treat!

CHAPTER FOURTEEN

The Pale Horse

While Hercule Poirot and Miss Marple are the most recognizable returning characters of Agatha Christie's canon, *The Pale Horse* (1961) includes another frequently revisited friend, Ariadne Oliver. As described on the Agatha Christie official website:

> Mrs. Oliver is a middle-aged woman and successful detective novelist, described as "handsome in a rather untidy fashion, with fine eyes, substantial shoulders, and a large quantity of rebellious grey hair with which she was continuously experimenting." She is feisty, quick to jump to conclusions (sometimes right, sometimes wrong), and strongly believes that Scotland Yard would be better run by a woman.[1]

Ariadne first appeared in a short story published in *Cosmopolitan* in 1932. Aside from *The Pale Horse*, Ariadne's other six novel appearances were all Poirot vehicles, as, other than Arthur Hastings, she is Poirot's closest friend. In interviews, Christie admits that Ariadne, often used as a comedic break from the macabre, was the character closest to her own personality:

> In Ariadne Oliver, author of such classics as *The Death in the Drain Pipe, The Affair of the Second Goldfish, and The Clue of the Candle Wax,* creator of the fabulous Finn Sven Hjerson, Agatha Christie offers a self-parody that is both comical and confident. Oliver is the kind of large, untidy, scatterbrained, middle-aged woman whom society tends to ridicule or ignore and whose professional aspirations were anathema to eminent Victorians like Charles Dickens, W. S. Gilbert, and Arthur Conan Doyle.[2]

When it comes to solving a murder, Ariadne functions much like Hastings. She is an aid, someone who might come up with the answer to a clue or two (because of her background in crime fiction), though in the case of *The Pale Horse*, she is the true "hero." Mark Easterbrook is a researcher of ancient cultures, who is pulled into the mystery of the Pale Horse organization, a sort of murder-for-hire plot. It's no wonder, as a rookie, that he needs Ariadne's and Inspector Lejeune's help!

Like a great many of Christie's novels, poison is used to dispatch the unfortunate victims in *The Pale Horse*. In this case, it is a pharmacist, Zachariah Osbourne, who is the culprit. And who would know the most about poisons but a pharmacist himself? Agatha Christie had an extensive knowledge of pharmacology due to research for her books as well as her own life experience as a pharmacist's apprentice.

> Though Christie was not formally trained as a pharmacist, she came to the world of pharmaceuticals as a volunteer nurse during World War I. While serving in Torquay Red Cross Hospital, she trained on the job and completed an exam that made her the equivalent of an assistant pharmacist. Christie resumed her duties in the pharmacy during World War II, performing thousands of hours of work in total.[3]

It was in the dispensary that Christie first thought of writing a detective story. Imagine if she'd never volunteered! Unlike nurse volunteers who were always on their feet, tending to ailing patients, Christie described many slow periods in the pharmacy when she had time to think about the mysteries she might like to write.

According to *The New Yorker*:

> Christie's fictions are profoundly shaped by the poisons that their characters skillfully employ. What's more, those characters enjoy relatively unfettered access to a range of exotic toxins, in a way that a would-be murderer could only dream of today. One begins to suspect that, among the many factors that gave us Christie's

> enormously popular novels, we must count the particular period for poisoning in which she lived.[4]

Christie became so intimately familiar with medicines, she recounted a time during World War I when she realized the pharmacist she worked under had made a grave mistake in calculating the amount of lethal drugs in a batch of suppositories: "I didn't like it, and what was I to do about it? Even if I suggested the dose was wrong, would he believe me? There was only one thing for it. Before the suppositories cooled, I tripped, lost my footing, upset the board on which they were reposing, and *trod on them* firmly."[5] This sort of physical humor laced with the morbid is so "Agatha Christie" at its core, and reminds us, too, of her young heroines like Bundle Brent.

This pharmacist, who would've likely killed patients with the suppositories had Christie not dropped them, is named "Mr. P." in her autobiography. With her trademark attention to human absurdity, she tells another story of Mr. P., half a century in the making:

> He was a strange man, Mr. P. One day, seeking perhaps to impress me, he took from his pocket a dark-colored lump and showed it to me, saying, "Know what this is?"
>
> "No," I said.
>
> "It's curare," he said. "Know about curare?"
>
> I said I had read about it.
>
> "Interesting stuff," he said, "very interesting. Taken by the mouth, it does you no harm at all. Enter the bloodstream, it paralyzes and kills you. It's what they use for arrow poison. Do you know why I carry it in my pocket?"
>
> "No," I said, "I haven't the slightest idea." It seemed to me an extremely foolish thing to do, but I didn't add that.
>
> "Well, you know," he said thoughtfully, "it makes me feel powerful."
>
> His memory remained with me so long that it was still there waiting when I first conceived the idea of writing my book *The Pale Horse*—and that must have been, I suppose, nearly fifty years later.[6]

We're grateful Christie used such a colorful character for inspiration, even if it was five decades later!

The *Agatha Christie's Marple* television adaptation of *The Pale Horse* alludes to *The Exorcist* (1973) in the opening sequence. As the priest leaves the victim, having given her last confession and provided him with a list of names, he makes his way down a steep staircase, under a streetlamp, and is murdered. Thankfully, he was able to mail the list of names and the Bible verse which talks about the pale horse off to Miss Marple.

The novel is considered to have saved at least two readers from thallium poisoning. What is it? Thallium is a compound that was used in rat and ant poisons up until 1975. It's soluble in water, can be absorbed through the skin, and has been called "poisoner's poison" because it's colorless, odorless, tasteless, slow acting, painful, and can appear like other illnesses.[7] There was a thallium craze in Australia in the 1950s in which several women killed their husbands, relatives, or friends with the poison. In the 1970s, a serial killer was caught after a doctor consulting with police recognized the symptoms of thallium poisoning after reading *The Pale Horse*.[8] In 1975, Agatha Christie received a letter from a woman in Latin America who discovered she was being poisoned by her husband after reading the book, and in 1977, a nineteen-month-old infant was found to have thallium poisoning after a London nurse recognized the signs for the same reason.[9] Reading pays off, friends!

Three modern witches appear in *The Pale Horse*. How did people feel about witchcraft during this time period? The 1950s featured some books, music, and movies that delved into witches that may have shaped public opinion. The most well-known were two books, both by Gerald Gardner, called *Witchcraft Today* (1954) and *The Meaning of Witchcraft* (1959). They are credited with launching a new religious and spiritual movement despite their faults of some incorrect history.[10] The big witch movie of the 1950s was *Bell Book & Candle*, while songs like "I Put a Spell on You" by Jalacy "Screamin' Jay" Hawkins is still associated with witches today. Some people are still calling for witches to be cast out. In a sermon by Pastor Greg Locke in February of 2022, he claimed to know the names of six witches and stated that three of them were in the audience watching him speak.[11] The church, located in Tennessee, "held a bonfire to burn what it termed demonic materials, including items related to the popular Harry Potter and Twilight series, Ouija boards, tarot cards, and items connected to the Freemasons."[12] This sounds familiar! Read more about the history of this in our book *The Science of Witchcraft* (2022).

The Benjamin Abbot house in Andover, Massachusetts. "In 1692, Benjamin Abbot accused Martha Carrier of cursing him after a boundary dispute, which led to his own illness and the loss of his cattle."[13]

When Miss Marple arrives at the Pale Horse inn, it features a pyre outside. The receptionist at check-in confirms that the inn commemorates the burning of witches every year. Do people still commemorate this practice today? Absolutely! It may come as no surprise that the Salem Witch Trials Memorial in Salem, Massachusetts, pays tribute to the twenty victims who lost their lives in 1692 with granite benches that list the names, dates, and method of their executions. Quite movingly, "visitors will note that the words—among them, 'God knows I am innocent'—are cut off in mid-sentence, representing lives cut short and indifference to the protestations of innocence."[14] A monument in Norway honors the ninety-one witch trial victims during the years of 1600–1692 in Vardø. According to *Northern Norway,* there were a number of acts that could constitute being a witch, including "poisoning food, casting spells on domesticated animals, causing disease and death amongst people and casting spells on people. Meetings with Satan and gatherings involving drinking, dancing, and card games on Domen Mountain outside Vardø were also frequent occurrences."[15] It may be surprising to learn that Scotland carried out five times more executions per capita during their witch trials than the European average. History professor Julian Goodare, co-director of the Survey of Scottish Witchcraft, said "between the first execution in 1479 and the last in 1727, at least 2,500 women and men were killed and thousands more were tortured or put on trial."[16] Although there is no memorial in Scotland yet, advocates for the project located the perfect spot in 2022 for a proposed co-therapy wellness and leisure park that will educate the public and honor the victims.[17]

Mr. Venables has polio in *The Pale Horse*, and some believe it to be a lie. What is the history of this condition? Ancient Egyptian paintings show otherwise healthy people with withered limbs and children using canes. This is believed to be the first proof that polio has existed for centuries. Currently, polio has fewer than one thousand cases per year and is preventable by vaccine. The symptoms include sore throat and fever but can progress to severe headache, neck stiffness, and paresthesia, a tingling and prickling of the skin.

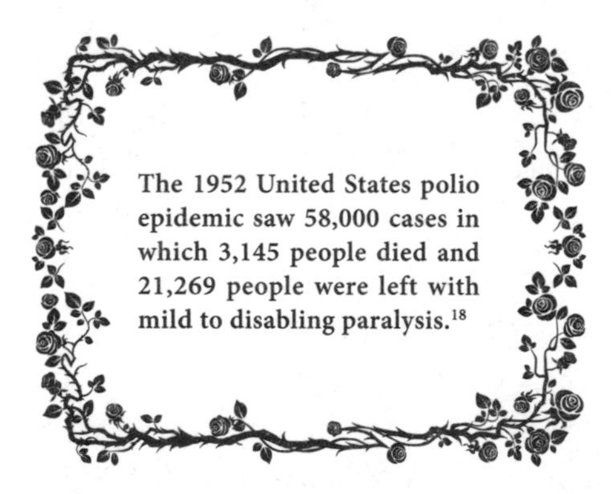

The 1952 United States polio epidemic saw 58,000 cases in which 3,145 people died and 21,269 people were left with mild to disabling paralysis.[18]

The clinical term for people who fake diseases or paralysis is "malingering," which is "the fabrication, feigning, or exaggeration of physical or psychological symptoms designed to achieve a desired outcome, such as relief from duty or work."[19] One of the most notable cases of malingering took place after a twenty-five-year-old woman named Desiree Jennings received a flu shot in 2010 and claimed to experience fever, joint pain, a British accent, memory loss, and stuttering. She discovered the condition of dystonia online that is characterized by involuntary muscle contractions. She read that some people are able to move normally by running, which she started to do almost immediately. "She also found out she could walk backwards, and even sideways, and that while doing so, her speech returned to normal."[20] Although some doctors consider her condition a hoax,[21] Jennings stood by her claims.

A dog in *The Pale Horse* has ringworm. What causes this and what is the treatment for it? Ringworm is a fungal infection of the skin that appears as lesions of hair loss in dogs and is typically circular in shape. The areas can appear throughout the dog's body and nails and are not usually itchy. Ringworm in dogs can be treated through topical creams, ointments, or shampoos and can sometimes be used in combination with oral medications.[22] Can humans be affected by ringworm? Yes, and it can appear on the skin as athlete's foot or jock itch. These can be treated with over-the-counter remedies, while ringworm on the scalp needs to be treated with prescription antifungal medication.[23]

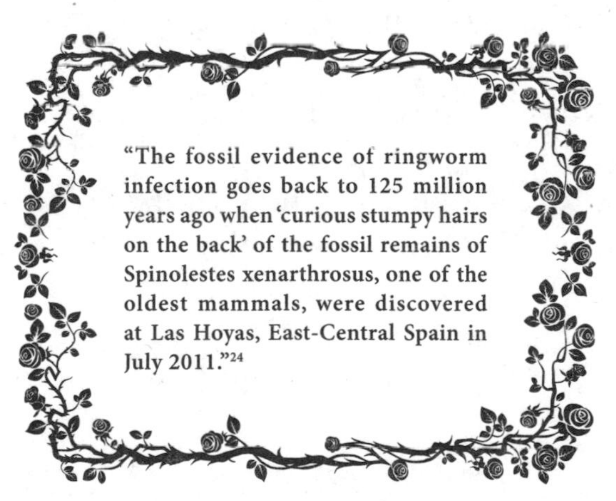

"The fossil evidence of ringworm infection goes back to 125 million years ago when 'curious stumpy hairs on the back' of the fossil remains of Spinolestes xenarthrosus, one of the oldest mammals, were discovered at Las Hoyas, East-Central Spain in July 2011."[24]

The victims' hair loss in *The Pale Horse* is attributed to thallium poisoning, but what else could account for it? According to the Mayo Clinic, family history, hormonal changes including pregnancy, childbirth, menopause, or thyroid problems can contribute to hair loss. Changes in medications, supplements, levels of stress, and even certain hairstyles can cause hair loss. It's important to monitor nutrition and realize that conditions like diabetes and lupus can be contributing factors.[25]

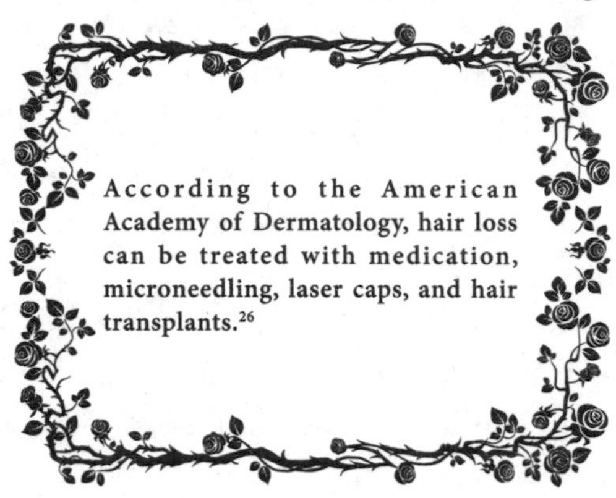

According to the American Academy of Dermatology, hair loss can be treated with medication, microneedling, laser caps, and hair transplants.[26]

The women of the Pale Horse organization go along with the ruse of "cursing" people. Using old superstitions combined with poison, people were convinced that the victims were dying by these witches' nefarious intentions. However, using information from consumer reports, the murderer was slowly killing his victims by replacing their everyday beauty items with poisoned ones. Miss Marple reveals that she, herself, wasn't poisoned because she keeps the labels of her products facing out. When she discovered one facing backward, she knew not to use it. She even left some extra hair in her brush to throw off the killer! This makes us think about how we place our day-to-day items and how we, too, could catch a culprit if need be. Leave it to Agatha Christie, once again, to inspire us in our own lives.

CHAPTER FIFTEEN

The Clocks

The beginning of *The Clocks* (1963) is a violent tableau of a stabbed corpse surrounded by six clocks, four of which are set at the same time. (And to think there are people who don't believe Agatha Christie wrote horror!) So starts yet another Hercule Poirot mystery, this one unique in that Poirot works to solve who is responsible using information alone, rather than any observation or interview tactics. We love our Belgian detective, but he sure can be a show-off! As Gillian Gill points out in *Agatha Christie: The Woman and her Mysteries*, there has been little change in Poirot from his introduction in the 1930s to three decades later in *The Clocks*. He still dyes the gray from his mustache, though has lost a bit of his zeal. "In 1963's *The Clocks* Poirot is a shade more reluctant to leave his apartment than in earlier days and no longer runs wildly around when inspired as he once did in the gardens of Styles Court."[1] Although his age seems to have roughly stayed the same, perhaps Poirot is reflecting the passage of time in the author herself? Whatever the reason, Poirot is more than happy to let the younger Colin Lamb do the proverbial heavy lifting. Later in life, Agatha Christie lamented how "I saw what a terrible mistake I had made starting with Hercule Poirot so *old*—I ought to have abandoned him after the first three or four books, and began again with someone much younger."[2] Instead, she allowed Poirot to live in a sort of stasis, aging extremely slowly or sometimes not at all, like many beloved fictional characters (think Bart Simpson). Some have even attempted to figure out Hercule Poirot's age with calculations. In their book, *The Agatha Christie Companion* (1984), authors Dennis Sanders and Len Lovallo determine that Poirot aged two years for every three calendar years. If this is the case, he would be a whopping one hundred and one years old in his last book, *Curtain* (1975). This math is correct if one assumes he was about sixty in *The Mysterious Affair at Styles*.[3]

One of our favorite moments in *The Clocks* is when Inspector Hardcastle (working with Poirot and Lamb) is depicted with trademark Christie absurdity:

> Inspector Hardcastle walked in manfully. Unfortunately for him he was one of those men who have cat allergy. As usually happens on these occasions all the cats immediately made for him. One jumped on his knee, another rubbed affectionately against his trousers. Detective Inspector Hardcastle, who was a brave man, set his lips and endured.[4]

Many cats have frequented Agatha Christie's novels, like the peculiarly named vicarage cat Tiglath Pileser in *A Murder Is Announced.* (Tiglath was named after a famous Assyrian king who Christie undoubtedly came to know about on her frequent trips to Egypt and the Middle East.) As we have steeped ourselves in all things Christie to write this book, we have come to realize how integral pets were, not just in her fiction, but also in her personal life. In her autobiography, Christie briefly mentions the arrival of a special family member: "There was one lack in our lives: a dog. Dear Joey had died while we were abroad, so we now purchased a wire-haired terrier puppy whom we named Peter. Peter, of course, became the life and soul of the family."[5] Peter is Christie's most widely known dog, as he was the inspiration for *Dumb Witness* (1937), in which Hercule Poirot realizes that the owner of "Bob" (Peter's fox terrier counterpart) has tried to blame her dog for setting a ball on the stairs and accidentally killing a wealthy spinster, when it was her niece, Bella, all along! Bob gets to go home with Arthur Hastings once the murder is solved, so it's rather a happy ending. In fact, *Dumb Witness* is even dedicated to her beloved dog: "To Dear Peter, most fruitful of friends and dearest of companions, a dog in a thousand."[6]

On the official Agatha Christie website, her grandson, Mathew, appears in a four-minute video, speaking on Christie's love for dogs. He starts by showing Christie's last Christmas card she sent out before her death. On the front is a sepia photograph of herself at age five, holding a scruffy terrier, her first dog, George Washington. Mathew goes on to describe how vital dogs still are in the family, in big part due to Agatha Christie. He even jokes about her last dog, Bingo, who "bit everyone" but was still loved. He maintains that "perhaps the one she loved best was a short-haired terrier named Peter."[7] He goes on to explain that this was because Peter was there for her when her marriage fell apart. She came back from her disappearance and was greeted by Peter first, making her feel like everything was going to be okay. We encourage you to watch this video and the beautiful photos and memories of Peter, as I (Meg) was nearly in tears!

Agatha Christie's love of dogs is evident, as her works are peppered with wily terriers. Her short story "Next to a Dog," first published in 1929 in *The Grand Magazine* (1905–1940) and later reprinted in 1971 in *The Golden Ball and Other Stories*, centers on a widow who loves her terrier so much, she'll do anything to keep him.

The aforementioned Manchester terrier, Bingo (who liked to bite, as Christie's grandson described), was immortalized in fiction, too. He appears as "Hannibal" in *Postern of Fate* (1973) as Tommy and Tuppence's faithful dog:

> Hannibal assists at several points in the book, including warning Tommy and Tuppence that there is a shooter hiding in the garden. He also gains a trophy for himself in coming away with a sample of the criminal's trousers. Further chasing and biting of the bad guys also crops up later in the book. Yet Hannibal acts as more than a guard dog in this story, as his desire to go into the churchyard, ignoring Tommy's instructions, leads to Tommy uncovering an important grave head stone.[8]

Agatha recounts:

> Many friends have said to me, "I never know when you write your books, because I've never seen you writing, or even seen you go away to write." I must behave rather as dogs do when they retire with a bone; they depart in a secretive manner and you do not see them again for an odd half hour. They return self-consciously with mud on their noses. I do much the same.[9]

There are no dogs to save the day in *The Clocks*. Hardcastle and Lamb interview suspects, while Hercule Poirot cracks the case even though he never goes to Wilbraham Crescent, the scene of the murder. He might be a show-off, but Hercule Poirot has a superior intellect . . . so we guess he's allowed a little showboating!

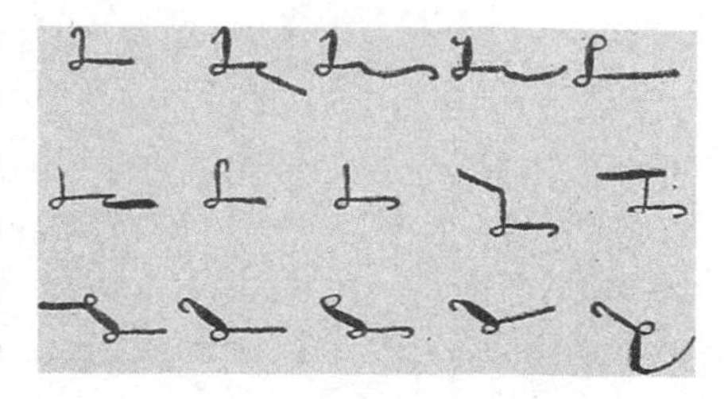

An example of shorthand writing.

The character Sheila Webb is a stenographer in *The Clocks*. What is the history behind this profession? Stenography comes from the Greek words *stenos*, meaning narrow, and *graphy*, meaning writing. Initially, shorthand was used and written by hand to document speech using symbols. I (Kelly) remember my mom, who was a medical transcriptionist, writing our annual Christmas lists in shorthand so my sister and I couldn't translate! She would tease, "would you like to see what we bought you for Christmas?" We would be *so* excited to catch a glimpse, only to have a nonsensical (to us) language shown on a sheet of paper. Eventually, she switched to a modern computer. Stenotype machines, which consist of twenty-two keys, are still used in courtrooms today.

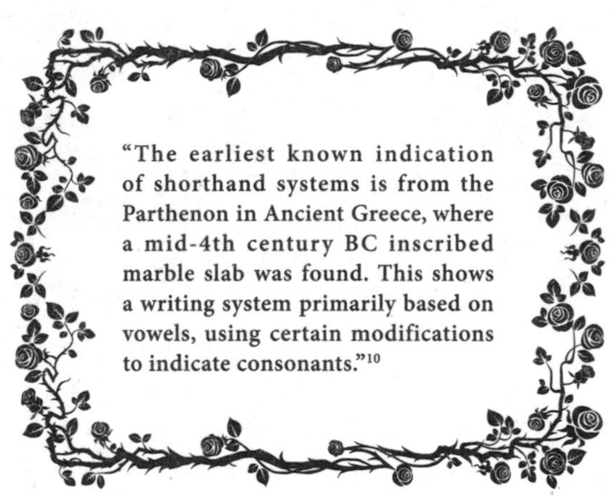

"The earliest known indication of shorthand systems is from the Parthenon in Ancient Greece, where a mid-4th century BC inscribed marble slab was found. This shows a writing system primarily based on vowels, using certain modifications to indicate consonants."[10]

Humans have used different instruments to measure time over the centuries. First, we used the sun to tell time with inventions such as sundials and shadow clocks. Another invention, candle clocks, determines the time of night according to the rate of the burn. Hourglasses became the primary way to tell time at sea and are still used today throughout the world. Clock towers came next in history, and one of the oldest ones, built by a monk in the fourteenth century, is still in existence today at London's Science Museum! The minute hand, pendulum clocks, and mechanical clocks came next in the following centuries and by 1878, standard time was invented.[11]

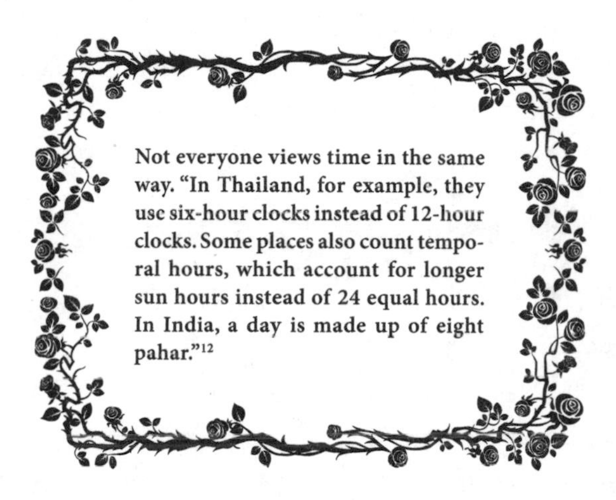

Not everyone views time in the same way. "In Thailand, for example, they use six-hour clocks instead of 12-hour clocks. Some places also count temporal hours, which account for longer sun hours instead of 24 equal hours. In India, a day is made up of eight pahar."[12]

One of my (Kelly) favorite memories is spending the night at my grandparents' house in the upstairs guest bedroom. Numerous clocks were present in the room, all ticking at different intervals. It was a calming, familiar sound, but chaos would ensue in the morning if I set multiple alarms and the clocks weren't exactly aligned in their time! Many of the clocks in *The Clocks* are running fast, set to 4:13 pm. What causes a clock to run too fast or slow? For the clocks in the time period in which this story was written, environmental factors could have been at play. The temperature or humidity could have affected the internal mechanisms, causing them to change the speed. Pieces could be attached incorrectly, there could be a lack of lubrication, or the balance in pieces might need to be adjusted.[13] To fix a digital clock that runs fast, try changing the batteries or plugging it into a different outlet. It's important to always consult an expert in clock repair, especially when dealing with a vintage or antique clock, to get an accurate diagnosis. Of course, in *The Clocks*, the multiple clocks stopped at 4:13 were nothing more than a red herring.

Braille is used to send coded messages in *The Clocks*. What is the history behind it? "Night writing" was invented in the 1800s when French soldier Charles Barbier developed a system for soldiers to safely write and read messages in the dark. Those who used lamps to write messages could be spotted by their enemies and killed. "Barbier based his 'night writing' system on a raised twelve-dot cell; two dots wide and six dots tall. Each dot or combination of dots within the cell represented a letter or a phonetic sound. The problem with the military code was that the human fingertip could not feel all the dots with one touch."[14] Louis Braille, who was blind himself, later adapted the night writing system to be six dots instead of twelve and by 1860, it reached the United States.

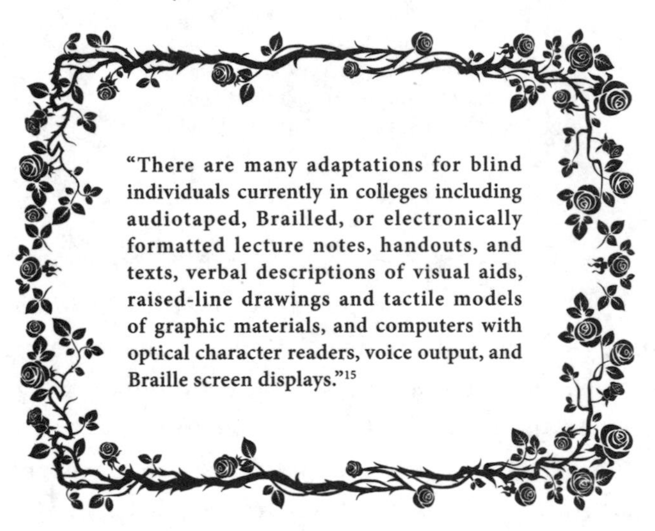
"There are many adaptations for blind individuals currently in colleges including audiotaped, Brailled, or electronically formatted lecture notes, handouts, and texts, verbal descriptions of visual aids, raised-line drawings and tactile models of graphic materials, and computers with optical character readers, voice output, and Braille screen displays."[15]

The Clocks not only inspired us to learn more about the various methods of communication used in this story, but also about our own perception of time. Agatha Christie always leaves us with a lot to think about in her brilliant mysteries!

CHAPTER SIXTEEN

By the Pricking of My Thumbs

"By the pricking of my thumbs,
Something wicked this way comes."[1]

Graduating from her habit of borrowing from nursery rhymes, Agatha Christie's *By the Pricking of My Thumbs* (1968) takes its name from Shakespeare's *Macbeth*. Several years earlier, Ray Bradbury was also inspired by the Bard of Avon, naming his coming-of-age horror novel after the memorable line, *Something Wicked This Way Comes* (1962).

By the Pricking of My Thumbs marks the return of beloved couple Tommy and Tuppence Beresford. First appearing in *The Secret Adversary* over forty years earlier as youthful lovers, they had last been seen in Christie's spy novel *N or M?* (1941). In the early 1940s, they were middle-aged and pleased to be pulled into the intrigue of World War II. *N or M?* made such a splash in England that Agatha Christie was investigated by MI5, the UK's domestic counterintelligence and security agency. Why?

> The answer, it can now be revealed, lay in the name of a character in her wartime novel *N or M?*, whom she called Major Bletchley. He appears in the book as a friend of Christie's pair of detectives, Tommy and Tuppence. In the book, published in 1941, N and M are the initials given to two of Hitler's agents as Tommy and Tuppence hunt for the enemy within. Major Bletchley comes across as a tedious former Indian army officer who claims to know the secrets of Britain's wartime efforts. Christie happened to be a close friend of Dilly Knox, one of the leading codebreakers at Bletchley

> Park. MI5 was concerned that the major's inside knowledge of the progress of the war was based on what the codebreakers knew about Hitler's plans. Had Christie mischievously named the character Bletchley because Knox told her what was going on there? When confronted, Christie responded, "Bletchley? My dear, I was stuck there on my way by train from Oxford to London and took revenge by giving the name to one of my least lovable characters."[2]

It's unbelievable to think that in the midst of World War II, MI5 used their precious resources to investigate Christie!

Over a quarter of a century later, Tommy and Tuppence returned in *By the Pricking of My Thumbs,* now in their sixties. Like Poirot and Miss Marple, they do not fade into a restful retirement, instead jumping into mayhem full force! Agatha Christie often said Tommy and Tuppence were two of her favorite characters, even though their novels were not her most highly praised or well known. Biographer Gillian Gill explores the Beresfords' appeal to Christie:

> The answer lies more in the relationship Christie created between her two sleuths than in their detective exploits. Superficially, Tommy and Tuppence have highly conventional middle-class lives, with conventionally segregated sex roles. In the 1929 *Partners in Crime*, for example, Tommy masquerades as head of a detective agency, while Tuppence is merely his secretary. Once Tuppence is pregnant, Tommy pursues a successful bureaucratic career while his wife devotes herself to the children and the house. The Beresfords, however, lead conventional lives only in the space between novels. The novels' actions, as opposed to the biographical link passages, show the Beresfords as a team of equals, not as boss and female sidekick. As the plot carries them into one death-defying situation after another, each relies on the other's competence and admires the other's special talents. Tommy provides the common sense, the brawn, and the professional contacts, while Tuppence provides the brainpower, the flair, and the audacity.[3]

Is there anything more romantic than a married couple fighting crime together? Christie didn't think so. Tommy and Tuppence bring forth thoughts of Agatha and Max, working together at archeological digs, relying on each other in difficult terrain. No wonder she felt a special connection with the aging Beresfords!

The dedication in *By the Pricking of My Thumbs* says it all: "This book is dedicated to the many readers in this and in other countries who write to me asking: 'What has happened to Tommy and Tuppence? What are they doing now?' My best wishes to you all, and I hope you will enjoy meeting Tommy and Tuppence again, years older, but with spirit unquenched!"[4]

There is an element to Agatha Christie's novels that we've not delved into yet, and it happens to be my (Meg) favorite. As a preteen, I noticed something different in many of Christie's books, usually on the first or second page. I would pull out her paperbacks at the library, finding interest in and examining these treasures, a clue to what macabre mischief the book would hold. They were maps, of course. Unlike the intricate ones I'd seen in fantasy novels, these were smaller scale, focusing on a mansion, or sometimes just one room where the murder occurred. In *Murder on the Orient Express*, for example, a map of the train's sleeping car is provided with each chamber marked with the suspect inside. This inclusion of crime scene maps in Christie's books adds another interactive layer to her mysteries.

Christie is not the first to incorporate these diagrams into her work, as it was a pairing of two art forms long before. In 2022, the Huntington Library, Art Museum, and Botanical Gardens in San Marino, California, held an exhibition, "Mapping Fiction," to honor such a rich and storied tradition. The Huntington Museum curator, Karla Nielsen, tells *The Guardian*:

> When Robert Lewis Stevenson's publisher attempted to bring out his 1886 novel *Kidnapped* without the author's map, Stevenson was outraged. Without the map, *Kidnapped* didn't function the way Stevenson wanted it to, he really wanted readers to be able to understand how this kidnapped character was being moved

> around. Having that topographic awareness as a reader gives you a level of control over the story that the protagonist doesn't have.[5]

As Kristel Autencio writes in her article, "Grounds for Murder: Maps and Floor Plans in Mystery Novels," Agatha Christie is not the only mystery author to utilize a map in spinning a murder tale. Dorothy L. Sayers provided a two-story floor plan of a hunting lodge in *Clouds of Witness* (1926), and Umberto Eco included a number of diagrams and maps of a fourteenth century abbey in *The Name of the Rose* (1980). "Mystery novelists have been using maps . . . early on, most notably in 1908 with Gaston Leroux's *The Mystery of the Yellow Room*. The narrator even breaks the fourth wall and says, 'The plan of the ground-floor only, sketched roughly, is what I here submit to the reader,' before giving the rundown of the crime scene."[6]

I was so taken with the use of Agatha Christie's maps, I included my own map in my first novel, *Her Dark Inheritance* (2018). By showing my fictional town of Willoughby in map form, I was adding in the world building. (And, thankfully, my mom is a talented artist who gave me a great discount.) It was a little nod to Agatha Christie, which I hope to do again.

Tuppence suffers a concussion after getting knocked out by a blow to her head while investigating the local church cemetery in *By the Pricking of My Thumbs*. What is the science behind concussions and how do they affect our brains? Our brains are extremely sensitive and impressionable, like the consistency of a slightly set Jell-O, and float unattached inside our skulls. Because of this, a concussion can cause temporary and sometimes permanent cognitive damage. The brain can be injured in three ways: focal impact, linear impact, and angular impact. The focal impact is where the brain was hit, while the linear impact occurs with no direct impact, such as whiplash. Angular impact occurs when the head is twisted suddenly and separates the brain from the spinal cord temporarily.[7] Signs of concussions include loss of consciousness, balance problems, amnesia, and vomiting while ongoing symptoms are likely headache, dizziness, sensitivity to light and noise, difficulties with attention, focus, and sleep.[8] It's important to seek medical treatment if you suspect you or someone else has experienced a concussion.

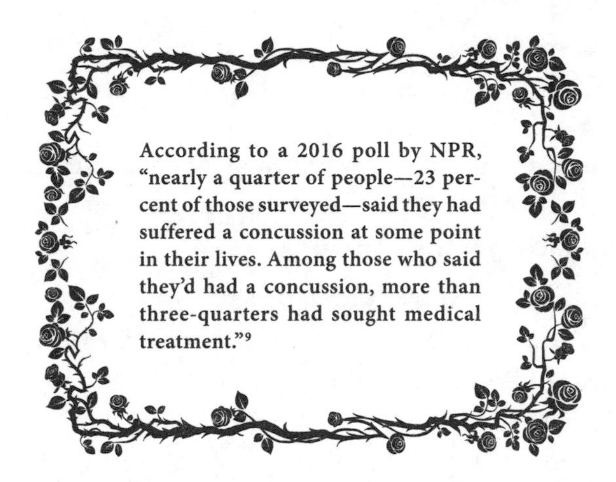

According to a 2016 poll by NPR, "nearly a quarter of people—23 percent of those surveyed—said they had suffered a concussion at some point in their lives. Among those who said they'd had a concussion, more than three-quarters had sought medical treatment."[9]

By the Pricking of My Thumbs contains a mysterious painting, a supposed haunted house, and a child's coffin filled with treasure. These sound like the makings of an elaborate escape room mystery! We had the opportunity to interview escape room designer Luke Moravec about how he creates mysteries that people can experience in real time.

Kelly: **"You've designed escape rooms in the past. Tell us about your history with it and the process of coming up with an initial theme or concept."**

According to World of Escapes, people enjoy escape rooms to fulfill their natural sense of curiosity, feel part of a team, and have fun playing a game.[10]

Luke Moravec: "I've been hosting and designing escape rooms for Solve Entertainment Escape Rooms in Duluth, Minnesota, for the past three years or so. Coming up with a theme for an escape room is not unlike coming up with an idea for a novel; figure out what interests you, what would be intriguing, find some compelling characters and a rich setting, and you're already out of the blocks. From there, understanding the tone of what you're going for is important. If you can imagine what you want players to be feeling while going through the experience, you've got all the foundation you need. The tone of a game is so important. In a nutshell, setting, characters and tone. Any sort of mystery, puzzle, or plot will blossom from those three categories."

Meg: **"How do you plan or write the story for each individual room/experience? Is it linear or do you have ideas for hints/props and incorporate them in?"**

Luke Moravec: "At Solve, the games are never entirely linear. Creating a game that can be experienced, played with, and solved in various, different ways is important. A large part of that is simply because when you have a team of, say, eight players in one room, you want them all to be playing. When a game is linear, there is a high likelihood of one or two of those eight players taking charge and leaving others to twiddle their thumbs. A good game will involve multiple points of attack and lots of moments for necessary teamwork. This also means doling out artifacts and items in an order that is a puzzle in and of itself. If an object that won't be used until the end is found in the first few minutes, it gives players something to muse about (or even forget about)! Much like a murder mystery, the vital clue might be on page one, but it doesn't make sense until the end. From a player perspective, it's very satisfying. From a designer/host perspective, it's thrilling to see eyes light up at the end of the game as the final objective is realized and someone shouts, 'Remember the scepter that was leaning against the doorframe when we came in?'

"When thinking about hints, props, and puzzles, everything should stay connected to the tone. Some escape room puzzles are really cool, but if you're designing an Egyptian-themed room, don't have a puzzle involving a 1930s-era cash register. This probably goes without saying (and this example is probably overly broad), but the tighter you can tie the puzzles to the tone, the more immersed your players will be."

Kelly: **"What are the pitfalls you run into when designing an escape room mystery?"**

Luke Moravec: "Mapping out the game on paper is crucial to avoiding pitfalls. Waiting until a game is built to troubleshoot it is not the correct strategy. Manage the game flow on paper, run through the different ways a team might explore the space and discover answers. If you notice (after drawing boxes and lines all over your notepad) that there is a superfluous key or a bit of information that's important but easy to overlook on the way to the end, you might want to figure out ways to angle players toward or away from (depending on the issue) solutions. Creating funnel points is also a good thing. As I've already said, having multiple things going on—to fill all those pairs of hands in the game—is a good thing, but ultimately the game will need to come to a conclusive point.

"Throughout the game, there will be moments like that too. Maybe the game has several rooms. If you solve everything in one room, sometimes there will be one lonely lock on a door that remains the only thing left to do. This isn't a bad thing; it's natural. It also allows the team to come together, share what they've learned, and let the story breathe for a moment. A good story in an escape is like a landed joke. Give everyone a second or two to think about it. It's just a chaotic mess if you're always plowing straight forward.

"Being too attached to certain elements that just aren't working can also be detrimental. Sometimes there are puzzles—good puzzles—that need to be eliminated. Maybe there are just too many or the tone isn't ringing true. Either way, sometimes you have to 'kill your darlings.' Inversely, don't cut something you feel is important to the story that you envision just because no one is seeing it. I wrote a game that was very close to my heart, and it was full of Easter eggs that I felt really enriched the story. For months, no one was noticing them. I started to think it was a waste of time. Then, one day, a young woman went through the game and pointed out some of my esoteric additions. It was a wonderful moment. Gentle foreshadowing can go a long way for some participants. It's not necessary for the enjoyment of the game, but when someone points it out, you know that the game just became that much more real (read, 'immersive')."

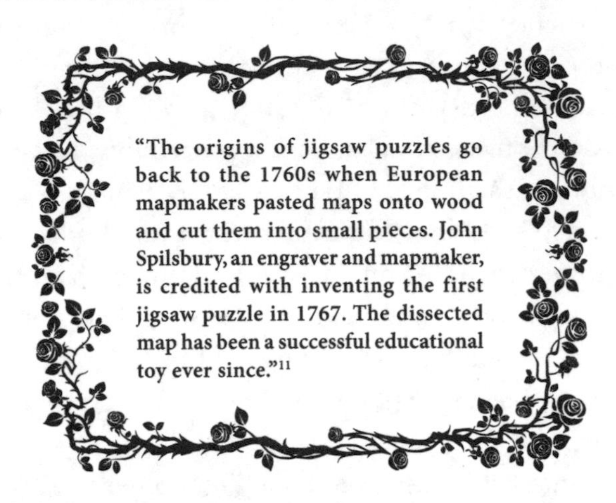

> **"The origins of jigsaw puzzles go back to the 1760s when European mapmakers pasted maps onto wood and cut them into small pieces. John Spilsbury, an engraver and mapmaker, is credited with inventing the first jigsaw puzzle in 1767. The dissected map has been a successful educational toy ever since."[11]**

Meg: **"What has surprised you the most as you've watched people go through the rooms you've designed?"**

Luke Moravec: "You can tell who the veteran escape room players are. They walk into a room and get down to business. Usually they collect suspicious items, talk clearly about things they observe, and rarely offer any whimsical comments about the environment or the goal. If you've got a good game on your hands, you'll see this change over the course of sixty minutes. By the end, an overly analytical team is often emotionally invested in the story, giggling, screaming, and less interested in how much time they have left than how much fun they just had."

Kelly: **"What has been your favorite room or experience you've designed or gone through yourself?"**

Luke Moravec: "Kuai Escape Rooms in Hawaii was a memorable experience. The preamble to our game suggested that several people had gone missing—last seen at a tiki lounge. Upon beginning our game, we were not only finding clues and solutions to puzzles, but also small snippets of the story here and there. By the end, the game had introduced several characters, a devious plot, and a truly unexpected final objective. That the designer allowed we, the players, to figure out the objective was welcome and, sadly, rare. One of my designs, 'The Babysitters,' really leans into that 'less is more' approach at the beginning. Some players don't like that at first; they want an objective outright. But that's like robbing the players of something else to discover. In 'The Babysitters,' players are told that they are responsible for babysitting for an hour—and they're not told much else. This really allows for open exploration and communication. Teams learn the beats of the story rather than having them stagnantly stated."

Meg: **"Do you have a favorite Agatha Christie book, movie, or play?"**

Luke Moravec: "I have fond memories of *The ABC Murders* (1936), and *Endless Night* (1967) is a sparse and elegant tale, but my favorite has always been *And Then There Were None.* Just as there is 'high-fantasy,' I think there needs to be a term called 'high-mystery.' That's what *And Then There Were None* is. It's absolutely everything that you could expect or want out of a good mystery and somehow the trope that it's become doesn't hinder enjoyment."

Thank you to Luke for the fascinating interview! Neither of us will ever go into an escape room again without looking at it differently.

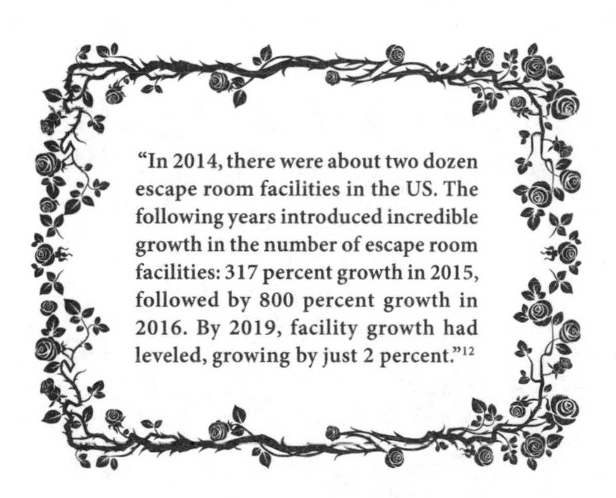

"In 2014, there were about two dozen escape room facilities in the US. The following years introduced incredible growth in the number of escape room facilities: 317 percent growth in 2015, followed by 800 percent growth in 2016. By 2019, facility growth had leveled, growing by just 2 percent."[12]

CHAPTER SEVENTEEN

Hallowe'en Party

It took a while, but at age seventy-nine, the Queen of Crime finally wrote a book, *Hallowe'en Party* (1969), about our favorite holiday! It is a natural fit, combining her trademark murder mystery with the most wonderfully frightful time of year.

As we've read through Agatha Christie's works, we've become intrigued by her dedications, and *Hallowe'en Party* is no less interesting: "To P. G. Wodehouse—whose books and stories have brightened my life for many years. Also, to show my pleasure in his having been kind enough to tell me he enjoyed my books."[1]

First, I (Meg) have to shamefully admit I didn't know much about P. G. Wodehouse. And second, it occurred to us that we didn't know a lot about Christie's social life, either. If she was having a Halloween party, would she invite Wodehouse? Were they friends?

In researching who P. G. Wodehouse was, we came across an article entitled "The Man Who Wrote the Most Perfect Sentences Ever Written." No wonder Agatha Christie held him in such high esteem! A British contemporary of Christie, Wodehouse was equally as prolific, writing ninety-six books in his impressive career. He died in 1975, less than a year before Christie.

> Some authors may want to expose the world's injustices, or elevate us with their psychological insights. Wodehouse, in his words, preferred to spread sweetness and light. Just look at those titles: *Nothing Serious, Laughing Gas, Joy in the Morning.* With every sparkling joke, every well-meaning and innocent character, every farcical tussle with angry swans and pet Pekingese, every utopian description of a stroll around the grounds of a pal's stately home or a flutter on the choir boys' hundred yards handicap at

a summer village fete, he wanted to whisk us far away from our worries. Writing about being a humourist in his autobiography *Over Seventy,* Wodehouse quoted two people in the Talmud who had earnt their place in Heaven: "We are merrymakers. When we see a person who is downhearted, we cheer him up."[2]

Biographers and literary critics have noted an obvious influence of Wodehouse on Christie. His books were already popular among England's wealthy class when Christie's *The Mysterious Affair at Styles* broke onto the scene. Charmingly, Wodehouse and Christie struck up a friendship in their later years, writing letters to each other like proper British authors. In 2018, letters between the two were published for the first time. They were donated by Wodehouse's estate to the British Library for exhibit, and, unsurprisingly, contain the sort of content two elderly friends would discuss: their health.

In a letter dated May 13, 1970, Dame Agatha lampoons the pretensions of clueless doctors who dole out antibiotics "until one of them does the trick or alternatively lays you in your coffin!" She also celebrates the loss of four stone "as a good result" of the heart attack she suffered in 1969. She writes: "Only nuisance has been that slimmer hips has resulted in a tendency to lose my skirt when I cross a street. Have to be held together by safety pins." The letters between the two highlight a shared admiration for each other's work—as well as a mutual distrust of the publishing business which made their names. Dame Agatha tells Wodehouse she is frequently infuriated by proof-readers who insist on "correcting" the grammar of her characters' dialogue. She says: "It really does enrage me, because few people I have ever met do talk grammatically and this includes myself." Dame Agatha also quickly tired of the numerous national celebrations in 1970 to mark her 80th birthday, writing: "Why should people congratulate one for being 80 years old—no damned merit in it."[3]

Agatha Christie is referred to above as "Dame" because she was given a "Dame hood" in 1971 as recognition by the British government of her service to literature. Queen Elizabeth II presented the honor. Dames and Knighthoods in literature are not doled out often. There are only five living literary Dames today. Christie's grandson, Mathew Prichard, said of his grandmother, "she was proud of the Dame hood but she got no more pleasure out of it than she did from the success of her books."[4]

I think it's safe to say that Agatha Christie would've invited her friend P. G. Wodehouse to a Halloween party. Although if it was anything like her book, he might've declined. You know, because of all the murder . . . .

Hallowe'en Party begins with a party thrown for children at the Elms School in Woodleigh Common. A game is played where the girls look into mirrors in hopes of seeing the faces of their future husbands. What is the tradition behind this? It turns out, there are several ways to find true love on Halloween! The mirror game has several recorded versions throughout history that include eating an apple, slice by slice, until your future mate appears in the mirror behind you to ask for the final piece. Another involves walking down a flight of stairs backward until you see your future husband in the mirror. (Good thing we didn't know about this practice before because it might have resulted in several falls down flights of stairs in our youth!) Another way to predict your future spouse? Peel an apple and toss the single strand over your shoulder. The shape of the letter it lands in will be the first initial of your sweetheart. According to fruit historian Joan Morgan, coauthor of *The New Book of Apples*, "early settlers of America brought with them European customs that stemmed from the age-old belief that apples were symbols of fertility."[5] Women secretly marked apples before dropping them into a tub of water for the bobbing for apples game. Men would predict their match based on the apple they bobbed for. Bobbing for apples of course comes into play during *Hallowe'en Party*, as Joyce is drowned in the bobbing tub.

The holiday of Halloween is ultimately known for donning creative and sometimes terrifying costumes. To learn more about the art of costume design, we had the chance to talk to Jolene Marie Richardson, a costume designer and fashion historian based in New York. She let us in on the fascinating world of costume design and helped us understand the process for film, television, and stage productions.

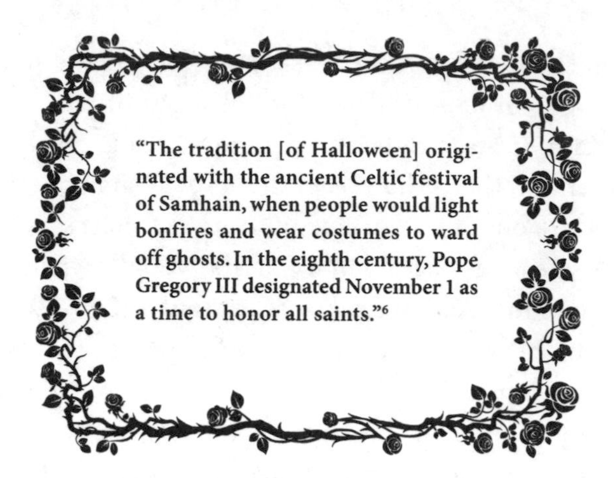

"The tradition [of Halloween] originated with the ancient Celtic festival of Samhain, when people would light bonfires and wear costumes to ward off ghosts. In the eighth century, Pope Gregory III designated November 1 as a time to honor all saints."[6]

Meg: **"First, could you tell us about how you became a costume designer? What inspired you?"**

Jolene Marie Richardson: "I was always a theatre kid growing up. I started acting when I was young and was in professional regional productions on Long Island in middle school through high school. I originally started college in the acting field but in an intro to costuming class we had to make a corset as our first project. I fell in love with making clothing. Sewing had been a part of my life, as my mother always made my Halloween costumes, and my grandmother my Christmas/Easter dresses, but I hadn't really dabbled much until college. It wasn't until grad school that I ever thought about being a designer, as I spent my undergrad career running and dressing shows (which I still do on occasion). I struggled to figure out what a designer really was my first few terms, then in my third term I had an incredible fashion history professor, Joan O'Clery, who showed me the power of the history of clothing. She honestly saved my butt in grad school. It clicked in my head what it meant to be a designer, and the importance of clothing. Now, when I was little, I should have realized that I would have done something in the clothing world; if an event required a costume, I was there! Even so far as when I was like five or six, I wanted to be a nurse, but quickly got away from that idea when I found out that they didn't wear the cute World War II uniforms anymore!"

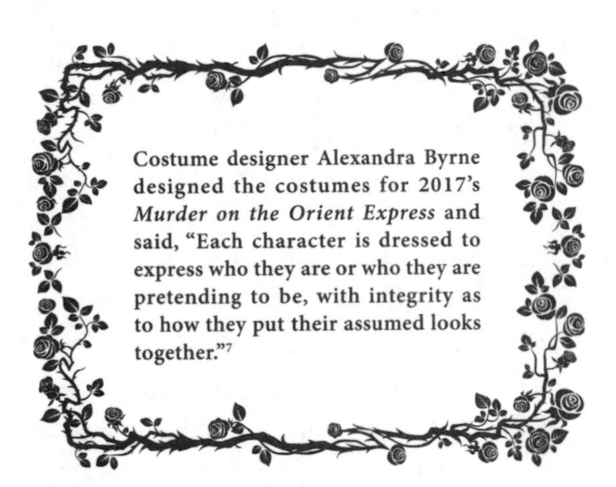

Costume designer Alexandra Byrne designed the costumes for 2017's *Murder on the Orient Express* and said, "Each character is dressed to express who they are or who they are pretending to be, with integrity as to how they put their assumed looks together."[7]

Kelly: **"You have worked in costuming in the horror genre. How does that differ from other genres?"**

Jolene Marie Richardson: "I wouldn't say that designing in horror is any different from designing a musical or a drama, but you are using different devices. In dramas, the clothing is telling you exactly who that person is . . . and what makes that character a unique person. In horror, usually you're subverting a lot of that outward symbolism. You want the audience to be thrown off or place the red herring onto someone else. Just like the timing of a jump scare, you want the attention to be somewhere else so that on the moment of reveal it makes a bigger impact. Now, something like *The Last Drive-In* (2018–) is a completely different animal, and I get to do things on that show I probably will never do anywhere else. I'm making things out of found objects with five minutes till shoot time, caking mud and guts on fabric, etc. I will also say that you see more risks being taken in design in the horror genre. They tend to be more fun; it reminds me a lot of my years designing musicals for the stage. Bold, bright, and a layer of whimsy!"

Meg: **"Agatha Christie uses clothing in her novels to convey particular character traits, like Poirot's fastidious nature. How does clothing in film work in a similar manner?"**

Jolene Marie Richardson: "Since film is a visual medium for very specific characters who know who they are, you really let the clothing do the talking. Poirot is a man of good standing, he's a detective, a businessman of sorts, and is very good at his job. He suits certainly reflect that. He deals with the crimes of the upper class and he himself is an upper-class man, so we are usually given him in a three-piece well-tailored suit. For a character like Poirot, the waistcoat is a key component. It adds an extra layer to his style that really separates him from his clients, almost in a protective way. Which comes back to the subverting theory, that he has nothing to hide but visually is telling us that by wearing all these layers he might be the one hiding. When we see our suspects, their garments in contrast tend to be more fluid and open yet they're the ones with secrets. From a practical perspective, he is a man of authority, and you can see that within the details of his clothing. Even the texture of his suits, they're solid but not too heavy, telling us that as a detective he's going to get to the bottom of the mystery, but he'll be understanding and hear all sides. Poirot is usually shown in subtle patterns; nothing too bold, shades of tans, blues, and grays. Professional colors but that also make him

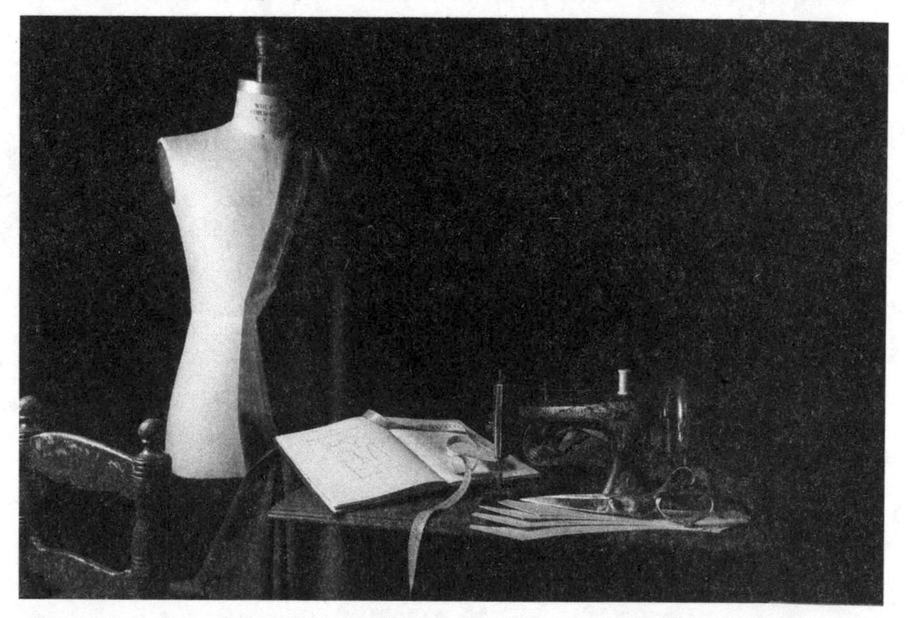

"The origin of fashion designing dates as far back as 1826. Charles Frederick Worth is believed to be the first fashion designer of the world, from 1826 to 1895."[8]

a relatable and trustworthy figure. You can see he has money, but he isn't flaunting his wealth or status. The accessories also add to who he is: cuff links, pocket watches, maybe a monocle, all inform us of his attention to detail and the eye for specific details he has when approaching a case."

Kelly: **"Could you speak to the film evolution of style over the decades of Hercule Poirot? Miss Marple? How do past and current Christie adaptations measure up?"**

Jolene Marie Richardson: "What I love about period films throughout different decades are each decade's interpretation of that specific moment in time, and each iteration really informs what time they were made in. When you approach a period piece, you're already working against the current because often the designer wasn't alive or there for that specific moment in time. So, you are starting from a secondhand account. Research helps a ton, but it still isn't the same as living in that moment. When you begin fittings for a project, there is a conversation that happens between the actor and the designer because each person has a specific view of who the character is, and you have to work together to come up with a vision that works for the actor since they're the ones being seen onscreen and have to wear the garments you've chosen. Sometimes there's coaxing involved but most actors are open to choices and styles that they normally wouldn't wear for the sake of the character. Men's suits in the last ninety years haven't changed too drastically, so it isn't hard to find something comfortable for your male actors.

"Where the influence of the moment in time when you are *making* the piece comes in would be in fabrics, colors, and textures. For one, textiles are made differently than they were in the 1920s; we have synthetic materials now (which were starting to become more widely used then but not to the extent that they are now). New adaptations, you can see the mood of even current climates through the costuming. Everything now is drenched in this dark realism. Designers tend to show Poirot in deeper shades; take

2017's *Murder on the Orient Express*. There's a darkness and a coldness to the costuming. Deep rich navy hues are used to invoke the mystery of the story but also reflect the climate of our world at large. We are exposed to more atrocities within the world because of social media and the way we consume our news. Our view of the world tends to be bleaker than adaptations from the 1980s or even the early 2000s."

Kelly: **"Are there clues in costume design in murder mysteries? Is there subtext in colors or fabrics that could help us discover the murderer?"**

Jolene Marie Richardson: "I think for designers you don't *want* people to outrightly guess, but you secretly plant pieces that once everything is revealed you can go back and say 'Oh, that's why she had ___.' You must tread the fine line between bearing that character's soul and hiding everything. It's like a symbolic tug-of-war. Disarm them so the scent is thrown elsewhere but plant subtleties for after the reveal."

Kelly: **"As we've been talking about Agatha Christie's *Hallowe'en Party* in this chapter, and you're a costume designer, we *must* ask—what have been your favorite Halloween costumes over the years?"**

Jolene Marie Richardson: "That's so much fun! Some of my favorites over the years have been Jessie the Cowgirl when I was seven or eight from *Toy Story 2* (1999) and Lily Munster from *The Munsters* (1964–1966) when I was nine. My mom made all my costumes growing up! She also made me a badass Harry Potter robe years before the movies came out, so she only had the *Harry Potter and the Philosopher's Stone* (1997) book cover to go off of! I was the green absinthe fairy from *Moulin Rouge* (2001) when I was seventeen and Miss Lizzie Borden! That was for a party I threw a few years back called 'dearly departed' where every guest had to come dressed as someone who was dead!"

Meg: **"Now, that sounds like our kind of party! Last, what are you currently working on and where can readers find you?"**

Jolene Marie Richardson: "Currently, I'm back to Broadway managing an off-Broadway show, *The Lucky Ones*. Design-wise, the newest season of *The Last Drive-In* premiered on Shudder on April 29, 2022. *Scare Package II* also came out in 2022; I was fortunate to design on the wraparound story with Aaron B. Kootz and the rest of the amazing Paperstreet Picture team. You can always catch my writing in the pages of *Fangoria* in print or online, and monthly I post on my personal blog *Hanging By A Thread*. I also co-host a podcast *To Dye For!* I try to keep busy!"

Thank you to Jolene for sharing her expertise with us! Make sure to follow her on social media and notice the costumes the next time you're watching a film, television, or stage production.

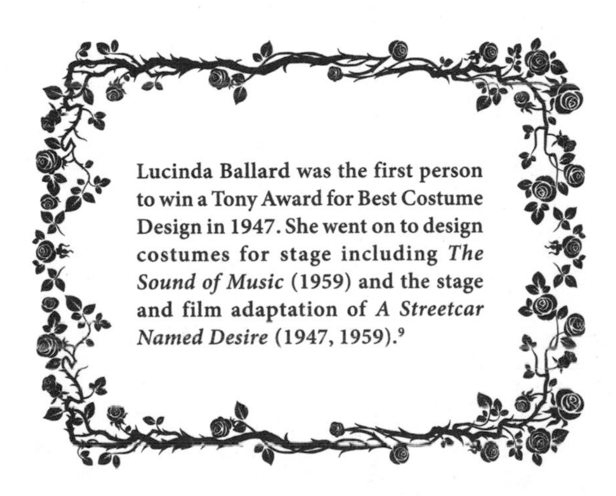

Lucinda Ballard was the first person to win a Tony Award for Best Costume Design in 1947. She went on to design costumes for stage including *The Sound of Music* (1959) and the stage and film adaptation of *A Streetcar Named Desire* (1947, 1959).[9]

CHAPTER EIGHTEEN

Elephants Can Remember

Nearly ten years ago I (Meg) picked up a tattered, hardcover edition of *Elephants Can Remember* (1972) at a used bookshop. I couldn't resist the seventies-esque pastel pinks and oranges on the cover. I took it with me on a vacation to Sarasota, Florida, and have lovely memories of sitting on our hotel room's balcony, lost in the world of Poirot and his good friend, Ariadne Oliver. If you have a good memory like an elephant, perhaps you'll remember that Ariadne is sort of Christie's fiction doppelgänger, an elderly author described in *Elephants Can Remember* as "a lucky woman who had established a happy knack of writing what quite a lot of people wanted to read."[1]

Agatha Christie has been a huge part of my reading life. I grab for her books when I crave a comfy read, something to spirit me away to England or Egypt when real life gets tricky. While writing this book, we've come to know the woman more, through biographies, her own words, and of course, through her fiction. So, I can't help but be defensive. As we've progressed through her oeuvre, it's been clear that her books have been given much poorer reviews in her later years. She was eighty-two when she wrote *Elephants Can Remember* and despite this (or perhaps because of this?), critics were harsh. They pointed out plot holes, and what they deemed sloppy syntax. We were surprised to learn that *Elephants Can Remember* was actually used in a 2009 study to show that Agatha Christie was suffering from Alzheimer's disease in the last years of her life. Those conducting the study believe this explains why her last books were given such egregious reviews. It's important to note that she was never diagnosed with dementia of any sort, so this is nearly impossible to prove.

> Academics at the University of Toronto studied a selection of Christie's novels written between the ages of twenty-eight and

> eighty-two, counting the numbers of different words, indefinite nouns, and phrases used in each. They found that the vocabulary size of the creator of Poirot and Miss Marple decreased sharply as she neared the end of her life, by 15 to 30 percent, while repetition of phrases and indefinite word usage (something, thing, anything) in her novels increased significantly. "We found statistically significant drops in vocabulary, and increases in repeated phrases and indefinite nouns in fifteen detective novels from *The Mysterious Affair at Styles* to *Postern of Fate*," said the academics, Dr Ian Lancashire from the English department and computer scientist Dr Graeme Hirst. "These language effects are recognized as symptoms of memory difficulties associated with Alzheimer's disease."[2]

In specific reference to *Elephants Can Remember*, the study found 30 percent fewer word types than in Christie's spy novel *Destination Unknown* (1954) that she wrote at age sixty-three. There were also 18 percent more repeated phrases.

When I first read *Elephants Can Remember* on that sunny balcony in Florida, I didn't notice any of this. I felt just as comforted by the characters and setting as I did reading her earlier novels. Applying this science to her words feels invasive, almost as if we are peering into her personal diary, but "signs of dementia in an author's work are not unprecedented. Peter Garrard, a cognitive scientist at University College London, found similar changes in the last book by British author Iris Murdoch. Shortly after publishing the book, Murdoch was diagnosed with Alzheimer's."[3]

This reflection on dementia in writing has also been studied outside the literary world. In 1990, the University of Minnesota published the "Nun Study." The University's David Snowdon:

> Wanted to look at aging over time, and decided to focus on sisters because they all had fairly similar histories and backgrounds. Most of them joined the School Sisters of Notre Dame congregation when they were eighteen, and all had abstained from smoking or drinking. So, Snowdon signed up 678 sisters, all over the age of

> seventy-five, from the order. All of the sisters agreed to donate a small part of their brains to the study after they died.[4]

It wasn't until after the study was underway that Snowdon and his fellow researchers discovered that most of the sisters had written an autobiographical account when they entered the convent, about fifty years earlier. Similar to how the University of Toronto applied science to Agatha Christie's fiction, Snowdon and his team analyzed the sisters' written words, leading them to an interesting conclusion. Those who displayed a higher vocabulary and more expression in their writing fifty years earlier (for example, "It was about a half hour before midnight between February 28 and 29 of the leap year 1912 when I began to live, and to die, as the third child of my mother, whose maiden name is Hilda Hoffman, and my father, Otto Schmidt")[5] were less likely to have been diagnosed with Alzheimer's disease or other forms of dementia, while those who wrote with less description (for example, "I was born in Eau Claire, Wisconsin, on May 24, 1913, and was baptized in St. James Church")[6] were more likely to suffer from dementia. In fact, those sisters who were in the bottom third of "idea density" in their autobiographies were sixty times more likely to develop Alzheimer's disease than those in the top third.

If we apply the logic of this "Nun Study" to Agatha Christie, there is a tragic irony, as she is one of the most well-known wordsmiths of the modern era, chock-full of vocabulary and "idea density." Yet despite this, she is *still* suspected of having Alzheimer's by scientists. However likely or unlikely something might be, disease can still strike anyone. Even the Queen of Crime.

One of her last novels, *Elephants Can Remember*, provides other clues, critics and scientists believe, that in her early eighties Christie was suffering from memory loss. Just the title alone, a nod toward the popular phrase "elephants don't forget," is pointed to by some as a public acknowledgement. Also, the plot centers on aging mystery author Ariadne, whose memory loss spurs on the investigation by her good friend, Poirot.

Ian Lancashire, who headed the Toronto study on Christie's novels, was shocked by the reduction of her vocabulary by one-fifth by the time she penned *Elephants Can Remember*, though he saw her sharing of her vulnerability, as well as her persistence in writing, admirable: "I began to see that Christie was heroic, still writing despite this handicap."[7] We, too, are humbled by Agatha Christie's continued commitment to weaving mysteries well past her "prime." Whether she was suffering from dementia or not, she certainly didn't let her advanced age hinder her love of what she did best.

In *Elephants Can Remember*, a body is found in a tub full of water that was used as a treatment for insanity. What is the history of this practice? Water has been seen to have healing properties throughout the centuries and can be traced back to the ancient Greeks. In the Middle Ages, manic people were repeatedly dipped into water to "cure" them. During the period of the Renaissance, a physician and chemist recommended patients be "fully immersed in cold water to the point of near-death in the hopes of 'killing the mad idea which caused mental derangement.'"[8] Hydrotherapy continued to be a popular method to treat mental illness in the early 1900s. Water was thought to be an effective treatment because, at different temperatures, it produces various reactions in the human body. Hydrotherapy was used throughout the early twentieth century and is still used today with more modern technology including pools, hot tubs, and physiotherapy tanks. The ailments modern hydrotherapy treats differ from the past and now include arthritis, amputations, and burns.[9]

"Regular winter swimming significantly decreased tension, fatigue, [etc.]"[10]

As the title of this story suggests, elephants are known for having good memories. What is the science behind memory in humans and in animals? Researchers have proven that elephants do, indeed, have exceptional memories that allow them to remember friends and enemies. A news story in 2022 featured an elephant who not only trampled and killed a woman but showed up to her funeral to trample her again.[11] She must have had a past with this elephant. Scientists have concluded that matriarchs in elephant herds can remember where they have previously found food and water, and remember those who hurt them.[12]

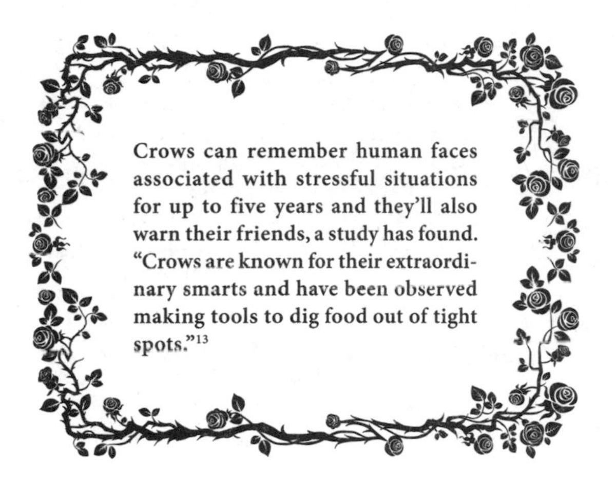

Crows can remember human faces associated with stressful situations for up to five years and they'll also warn their friends, a study has found. "Crows are known for their extraordinary smarts and have been observed making tools to dig food out of tight spots."[13]

Dogs may also remember past experiences. When I (Kelly) adopted a rescue dog from a local shelter, we were given a full psychiatric evaluation about his past. Apparently, our dog's brother was a bully and constantly told Bucky, our dog, that the sky was falling, and he believed it. It wasn't until Bucky's second month with us that he barked for the first time and started to show some confidence. Screw you, Norman! (Bucky's brother). What do scientists say about this? Dogs do have memories and may remember more than we think. According to *Smithsonian Magazine*:

> There are two forms of "explicit memory," which is the kind of memory you use when intentionally recalling a piece of information. The first is semantic memory, which you use to recall information you've consciously learned or memorized. The second is episodic memory, which you use to remember everyday

experiences and events that your mind encodes without conscious memorization.[14]

Episodic memory, which was defined by psychologist Endel Tulving in 1972, is related to self-awareness, so dogs' past memories could affect their behavior.

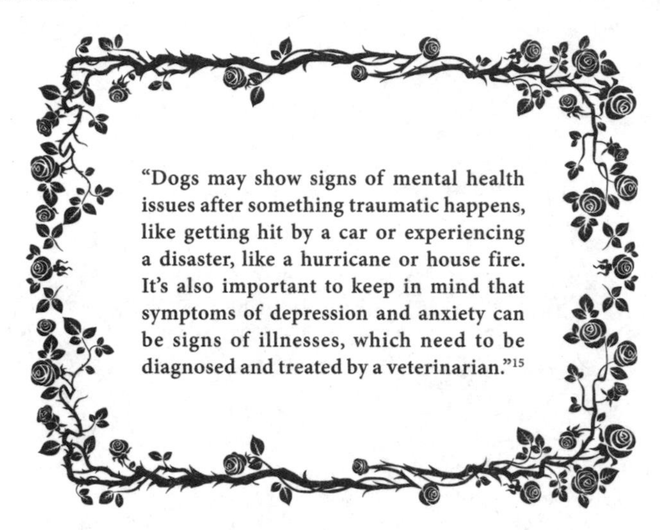

"Dogs may show signs of mental health issues after something traumatic happens, like getting hit by a car or experiencing a disaster, like a hurricane or house fire. It's also important to keep in mind that symptoms of depression and anxiety can be signs of illnesses, which need to be diagnosed and treated by a veterinarian."[15]

What should you do if you think your dog may be suffering from mental health issues due to traumatic memories from the past? The ASPCA (American Society for the Prevention of Cruelty to Animals) recommends seeing a veterinarian who is specially trained in animal behavior, maintaining a daily routine, reducing stressors in your pet's life, leaving a television or radio on for your pet when you're away, playing with them, and considering your own mental health state and how it may be affecting your pet.[16]

Poirot is investigating a murder suicide from twenty-five years ago. How possible is it to solve old cases? According to NBC News, "nationally, the solved-murder rate has fallen from 79 percent in 1976 to 69 percent in 2019. While the solved rate for white victims has increased to 81 percent in 2019, it has fallen to 59 percent for Black victims."[17] It is far more possible to solve cold cases in the current era due to the advances in DNA technologies, but in the decades that Agatha Christie wrote her stories, it was a bit less likely.

Even though people speculate that Agatha Christie may have been suffering from late-onset dementia, *Elephants Can Remember* is still a clever mystery that kept us on our toes!

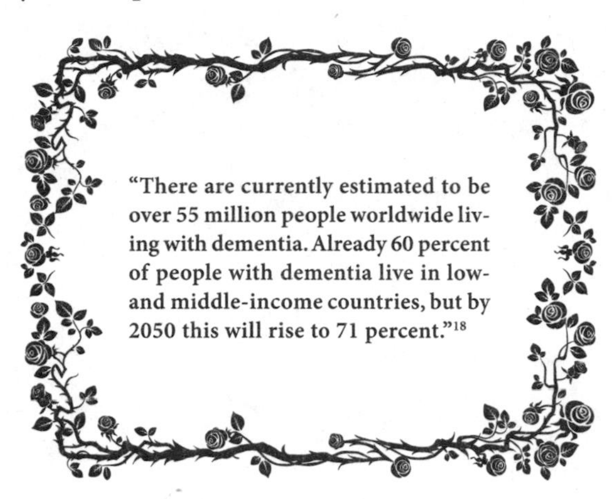

"There are currently estimated to be over 55 million people worldwide living with dementia. Already 60 percent of people with dementia live in low- and middle-income countries, but by 2050 this will rise to 71 percent."[18]

CHAPTER NINETEEN

Curtain

We have come to the end of Hercule Poirot's lengthy and impressive journey. Sort of. While *Curtain* (1975) may be the last case for Poirot, he has managed to transcend the boundaries of literature. To this day, just like his creator, he still looms in the zeitgeist, appearing in all reaches of media.

Curtain was the last case for Christie, too. (*Sleeping Murder* was published posthumously in 1976 but had been written decades earlier.) This end to both Agatha Christie and Hercule Poirot in one novel marks a poetic conclusion. Throughout this book, we've pointed to Christie's complicated relationship with her most famous character. In later years, she focused more on other characters rather than her persnickety Belgian detective. "It was during the postwar period that Miss Marple began to attract more and more popular attention. By this time Poirot was in great disfavor with Christie: she had killed him off in *Curtain*, one of the wartime manuscripts that were prudently laid away in a vault, and she cheerfully eliminated him from the 1951 stage version of *The Hollow*."[1]

Like *Sleeping Murder*, *Curtain* was first drafted twenty years before its release, allowing Poirot to live a bit longer before he was snuffed out by the woman who created him. In her autobiography, Agatha Christie shows a host of emotions toward Poirot. It was as though she were married to him, and not always happily: "I was by this time so stuck with Poirot that I realized I was going to have him with me for life."[2] There was even a semibiographical film that aired in Britain, *Agatha and the Midnight Murders* (2020), that highlights Christie's tumultuous regard for this character. Screenwriter Tom Dalton explains his approach to *Agatha and the Midnight Murders*:

> Christie wanted to write other characters, other settings within the mystery world, but here's Poirot just paying her bills over and over

> again, and I think this was part of the starting point for *Midnight Murders*, because she—in the previous eight years or so, she had written ten, eleven Poirot novels, it's a huge volume of work about one man. And so she was fed up with him, and crucially . . . she had made the decision to kill him off.[3]

It's not hard to believe that Christie and her publishers decided against publishing *Curtain* in the 1950s, as Poirot was still at the height of his popularity. Waiting until the end of Christie's life seems appropriate, as she knew all along that Poirot would be one of her greatest characters, her most recognizable contribution to detective fiction, whether she liked it or not!

Curtain is a befitting last case, as Poirot is past his prime and gets to go out with a bang (literally and figuratively). With a touch of welcomed nostalgia, Christie brought back Arthur Hastings for *Curtain*, who hadn't appeared with Poirot since *Dumb Witness* in 1937.

The foe, Stephen Norton, in *Curtain* is formidable, the sort of manipulative, Machiavellian sort that only Poirot, even elderly and suffering from a bad heart, can defeat. Instead of simply observing and using his "little gray cells," Poirot takes it upon himself to vanquish Norton. He murders him, and, because he is an expert in these things, masterfully covers up his crime. He dies soon after, content in knowing that he has gotten rid of a horrible man. This act of murder may sound shocking, but in novels like *Murder in Retrospect* and *Murder on the Orient Express*, we've witnessed how Poirot often usurps the typical route of justice for a more poetic form of what he believes is just. As longtime readers of Poirot's exploits, we feel his death is an appropriate one that further perpetuates his legend!

In 2013, David Suchet portrayed Poirot in the last episode of *Agatha Christie's Poirot* in an adaption of *Curtain*. After twenty-five years of portraying Poirot, Suchet was naturally swept up in a flurry of emotions:

> There is a complete strange mixture of . . . I now have to say goodbye, because I've done it, and then euphoria for exactly the same reason—I've done it! So it's a mixture of things. The

> predominant emotion is celebration, that actually it is time for him to go. The book was written; it's been on our bookshelves since 1975. Anybody could have read it, so nobody should be really surprised, especially the fans of Agatha Christie. She felt it was time, and having done all the stories, it is time [for me too]. So I'm very pleased to have done it—what a thing to leave behind, and personally, what a thing to have done.[4]

Saying goodbye is never easy, to a fictional character like Hercule Poirot, and especially to Agatha Christie herself.

I (Meg) recently read *Rock, Paper, Scissors* (2021) by bestselling British author Alice Feeney, a suspenseful book packed with mysterious deeds and life-changing secrets. It's the sort of book that probably wouldn't exist if Agatha Christie hadn't paved the way. As I was immersed in the story, I was met with a reference to Christie herself, as the fictional horror author in *Rock, Paper, Scissors*, Henry Winter, claims to own Agatha Christie's writing desk. Even in a break from working on this book, Christie finds me! She is ubiquitous and I love it.

This mention in such a freshly written novel made me wonder if her ephemera was actually out in the world. The first place I checked out was collectingchristie.com. There is an up-to-date blog on all Christie-related sales on eBay, as well as onsite auctions. They have quite a lot of information on first editions, like a rare copy, a bit rough around its edges, of *The Secret of Chimneys* (1925). It sold for $929 in March 2022. Items that Christie autographed seem to be worth even more. For example, a letter from 1971 that she typed and then signed to a friend named Mrs. Hastings-Wells. Although the content of the letter is quite innocuous (talk of cherry pie and chocolate coffee creams), it went for over $1,200.

In March 2022, a private seller auctioned off items they had amassed from a sale at Christie's Greenway Estate. Before the auction, a local paper in Sussex detailed what Christie collectors could bid on: "Dozens of items of furniture in the sale include a mahogany table, candlesticks, a tea service, a wooden trunk, several dictionaries, a desktop pad and writing ink. A five-inch tall Tibetan gilt copper alloy figure of Amitayus, dating from the 16th or 17th century, is valued at £2,000."[5]

If you're curious, Amitayus is one of two Buddhas in the Himalayan tradition. It "is typically depicted in the apparitional Buddha form (sambhogakaya), red in color, wearing a crown and jewels, holding a long life vase, above the two hands placed in a gesture of meditation."[6]

All these auctions, and even newly released goodies in honor of Agatha Christie, make us wonder what we need to save up for . . . .

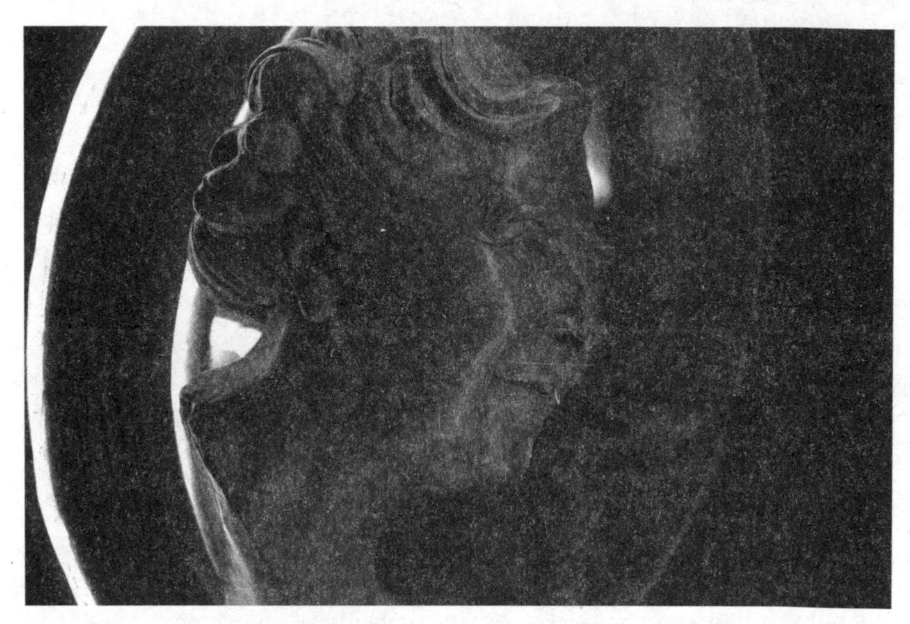

The Agatha Christie Memorial, located at the intersection of Cranbourn Street and Great Newport Street in London.

Poirot is elderly and arthritic in his last case. It's difficult to witness this fictional character go through the inevitable aging process, but it's a choice Agatha Christie decided to make. How does arthritis affect the body? The main symptoms are swelling and stiffness in the body, which typically worsen with age. Osteoarthritis causes cartilage to break down while rheumatoid arthritis specifically attacks joints. According to the CDC, nearly 24 percent of adults in the United States suffer from arthritis and it is the leading cause of work disability.[7] Arthritis can be treated with drugs, physical therapy, and in extreme cases surgery.

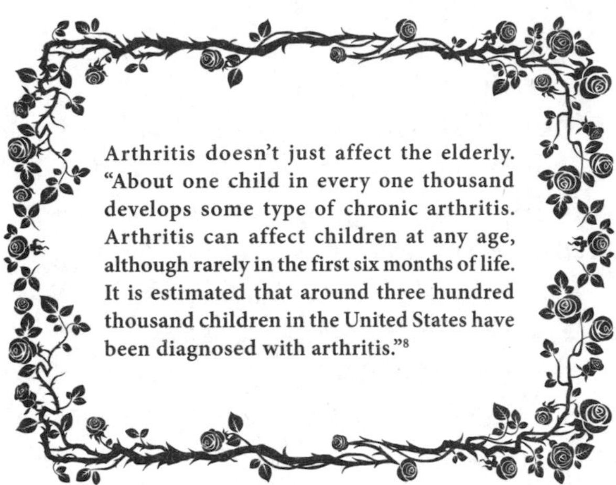
Arthritis doesn't just affect the elderly. "About one child in every one thousand develops some type of chronic arthritis. Arthritis can affect children at any age, although rarely in the first six months of life. It is estimated that around three hundred thousand children in the United States have been diagnosed with arthritis."[8]

Murder is committed by morphine in this story. What is the history behind this powerful drug? Morphine is derived from opium and was used in the mid-1800s to treat pain, tuberculosis, and "female problems" which included menstrual cramps, anxiety, and even morning sickness.[9] Morphine was also used prominently during the Civil War, and this contributed to a national addiction epidemic affecting one out of every two hundred Americans. What does a morphine overdose look like? Those going through an overdose may be agitated, have a change in their heart rate, feel sleepy, experience hallucinations, and show signs of confusion. Although morphine is used to murder in *Curtain*, accidental opioid deaths continue to be a growing problem in the United States. For heroin specifically, deaths "rose from 1,960 in 1999 to 15,469 in 2016. Since 2016, the number of deaths has trended down with 13,165 deaths reported in 2020."[10]

Morphine is often used when a person is in advanced stages of a terminal illness. "Research suggests that using opioids to treat pain or shortness of breath near the end of life may help a person live a bit longer."[11]

Another character is poisoned with physostigmine sulphate, an extract from the Calabar bean. This drug has a fascinating and terrifying history. Several cultures would use the bean, or the milk derived from crushing it and combining it with water, in trials as a method of detecting guilt. If a person was given the mixture and vomited, they were considered innocent. If they became more ill than just vomiting or ended up dying after ingesting the bean, they were considered guilty![12] Despite this horrific history, the bean was studied by toxicologists, and they discovered its medicinal properties. The drug physostigmine is used in medicine today to treat glaucoma, delayed gastric emptying, and has been shown to improve long-term memory.[13]

Norton has perfected the art of manipulating others to murder, like Iago in *Othello*. What is the science behind manipulation? Manipulators, by their nature, need to assert power and control over others. They tend to avoid taking responsibility for their actions and instead will blame and guilt others into thinking they're in the wrong. People who are manipulative are aggressive communicators who use anger to gaslight their victims. What are the effects of being manipulated? Those who have been in these types of relationships are often confused, question themselves, and feel shame and guilt over the situation they find themselves in. People who have been manipulated will avoid eye contact, display anxious tendencies, and can even go so far as to isolate themselves from others as their feelings of depression and isolation grow. Know that whichever side of a manipulative relationship you're on, it's possible to change.

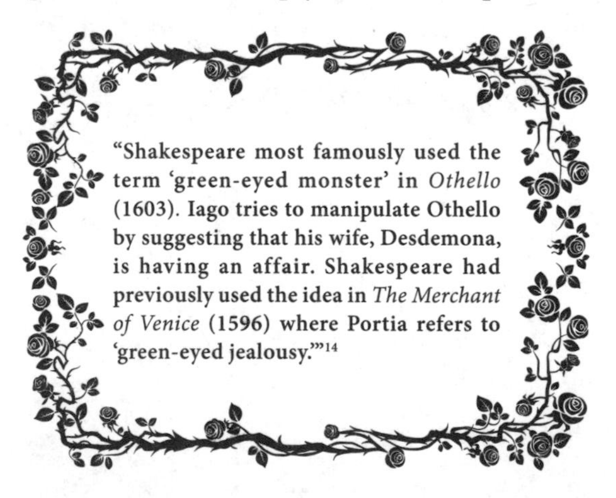

"Shakespeare most famously used the term 'green-eyed monster' in *Othello* (1603). Iago tries to manipulate Othello by suggesting that his wife, Desdemona, is having an affair. Shakespeare had previously used the idea in *The Merchant of Venice* (1596) where Portia refers to 'green-eyed jealousy.'"[14]

Poirot dies of a heart attack in *Curtain* after shooting Norton, an act he is not proud of but one that will ultimately prevent more murders. According to Agatha Christie's website, "The reception of Poirot's death was international, even earning him an obituary in *The New York Times*; he is still the only fictional character to have received such an honor."[15] We will truly miss this legendary character but, thankfully, he lives on in Christie's stories and the many iterations of Poirot in film and television.

CHAPTER TWENTY
Sleeping Murder

"Miss Marple insinuated herself so quietly into my life
that I hardly noticed her arrival."[1]

We now come to Agatha Christie's final published novel, *Sleeping Murder* (1976). As described in the previous chapter, Christie wrote *Sleeping Murder* during World War II, keeping it away from prying eyes for thirty years. In her autobiography, Christie describes the genesis of her last novel:

> I had written an extra two books during the first years of the war. This was in anticipation of my being killed in the raids, which seemed to be in the highest degree likely as I was working in London. One was for Rosalind, which I wrote first—a book with Hercule Poirot in it—and the other was for Max—with Miss Marple in it. Those two books, when written, were put in the vaults of a bank, and were made over formally by deed of gift to Rosalind and Max. They were, I gather, heavily insured against destruction.[2]

Like the author herself, *Curtain* and *Sleeping Murder* survived the bombing raids on London. *Curtain* is a novel of finality, the last case of Hercule Poirot, and a telling of his dramatic death, while *Sleeping Murder* finds Miss Marple very much alive at the end. If one were to read the Miss Marple novels in order of publication, they would be confused how she seems younger five years later in *Sleeping Murder* (in which she is down on her knees pulling weeds) than she is in *Nemesis* (1971). In *Nemesis*, the last written Miss Marple novel, the frail spinster avoids gardening on suggestion of her doctor. Oh . . . and it's set in 1944!

It's hard to know if Agatha Christie originally intended for *Sleeping Murder* to be Miss Marple's last case. She did make the decision to have it published in 1976, when she knew she would no longer write novels. Christie died about nine months before the book's release, seemingly content that it was her final contribution to a record-shattering career.

As we wrapped up our research on an incredible life, we found ourselves curious about *how* Agatha Christie wrote. (You know, just in case we, or you, wanted to become the next great author of detective fiction.) According to the official Agatha Christie website, she had to have paper handy: "She made endless notes in dozens of notebooks, jotting down erratic ideas and potential plots and characters as they came to her. 'I usually have about half a dozen (notebooks) on hand and I used to make notes in them of ideas that struck me, or about some poison or drug, or a clever little bit of swindling that I had read about in the paper.'"[3] Seventy-three of these notebooks full of Christie's thoughts were studied by author John Curran, then documented in his book *Agatha Christie's Secret Notebooks: Fifty Years of Mysteries in the Making* (2011).

Christie would begin a writing project with a lot of plotting and thinking before she put pen to paper. She was known to take long walks in Dartmoor, sometimes speaking dialogue aloud when no one was in earshot. As for her characters, Christie was an observer. Like many artists:

> [Christie] would observe people in restaurants and social gatherings as a starting point of creating her characters, jotting down their mannerisms and phrases. She had a strict rule about not using recognizable real people and felt strongly that the writer must always "make up something for yourself about them." She once said that the only time she tried to put a real person who she knew well into a book, it wasn't a success.[4]

Believe it or not, Agatha Christie didn't have a room dedicated for writing until late into her career. She often sat at a table in busy parts of the house with her favorite Remington Victor T portable typewriter.

She wasn't known to lock herself away to write, though she admits to doing this occasionally:

> Part of the secret of her astounding productivity was that she usually worked on at least two books at the same time. She also tried dictating to her secretary, Carlotta Fisher, but felt much happier writing in longhand and then typing it out, as this helped her keeping to the point. In her later years, after she broke her "writing wrist" she also used a Grundig Memorette dictaphone.[5]

Carlotta, sometimes known as "Carlo," was not only Christie's secretary, but also a longtime friend. She had first been hired to attend to Christie's mother, then became indispensable as Agatha Christie's career flourished.

A Google search of Christie's beloved writing tools brings a glimmer of what her process might've been like. We saw quite a few Remington T portable typewriters. They seemed common, as there were a number of antique ones for sale, ranging from $80 to $600, depending on their condition. Looking at a pristine one from 1936 nearly transported us to Devonshire, imagining the typing sound as Agatha Christie wrote of manners and murder. The Grundig Memorette dictaphone was a "newer" technology, first coming to market in 1960. It was enormous, though advertised as portable because it came inside a leather satchel. We'd be worried for fragile, elderly Christie carrying such a tech dinosaur on her shoulder! The machine comes with a boxy microphone that reminds us current writers how lucky we are to live in the era of laptops and iPhones.

Despite the fact that Agatha Christie is in the *Guinness Book of World Records* as the bestselling fiction writer of all time, she did not have a writing routine. No strict wake-up time. No late-night word counts. She found time to write when she could: through motherhood, marriage, friendships, moves, and even archeological digs to the Far East. We don't know about you, but this makes us feel so much more hopeful as writers!

Everything we uncover about Agatha Christie makes us more in awe of her talent and persistence. In the epilogue of her autobiography (written when she was seventy-five), Christie looked to the future. It's one of my (Meg) favorite quotes, as I feel it sums her up perfectly:

> I shall probably live to be ninety-three, drive everyone mad by being unable to hear what they say to me, complain bitterly of the latest scientific hearing aids, ask innumerable questions, immediately forget the answers and ask the same questions again. I shall quarrel violently with some patient nurse-attendant and accuse her of poisoning me, or walk out of the latest establishment for genteel old ladies, causing endless trouble to my suffering family . . . Until then, while I'm still comfortably waiting in Death's antechamber, I am enjoying myself.[6]

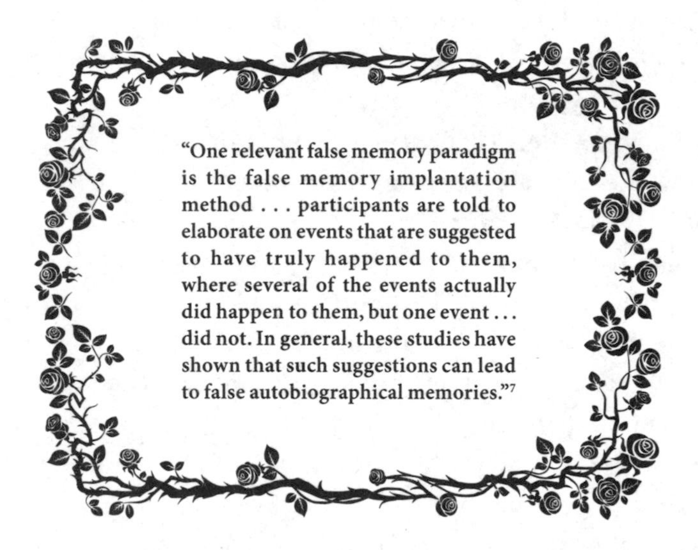

Repressed memories from childhood resurface for Gwenda in *Sleeping Murder*. How is this possible? Sigmund Freud was the first to postulate the theory that people possibly repress negative memories and they can come up later again in life. This theory is largely unproven and controversial. A Northwestern study in 2015 found that stressful or overwhelming experiences may trigger something in our brains. "A process known as state-dependent learning is believed to contribute to the formation of memories that are inaccessible to normal consciousness. Thus, memories formed in a particular mood, arousal or drug-induced state can best be retrieved when the brain is back in that state."[8] This study went on to experiment on mice, rerouting the processing of stress-related memories, and found that the mice didn't remember stress-related activities until they were returned to the state they were in when the stressor initially took place.

How could this be applied to humans? There are recent studies on how drugs can "rewire" the human brain and allow us to be more accepting of our situations in life. One example is the use of ayahuasca:

> [A] brew that combines the leaves of the Psychotria viridis plant boiled together with the Banisteriopsis caapi vine. Neither of these components has any power on their own, but together they are psychedelic. The ayahuasca brew contains DMT, a psychedelic chemical structurally similar to serotonin, and which also occurs endogenously in the human brain.[9]

This practice has been shown to reduce anxiety, depression, and social phobia in people as well as to treat some addictions. The experience may not be entirely pleasant, though. Many people report vomiting violently after ingesting the brew and experiencing hallucinations and tremors. Before embarking on any journey with drugs, consult your primary care doctor to determine what is safe and right for you.

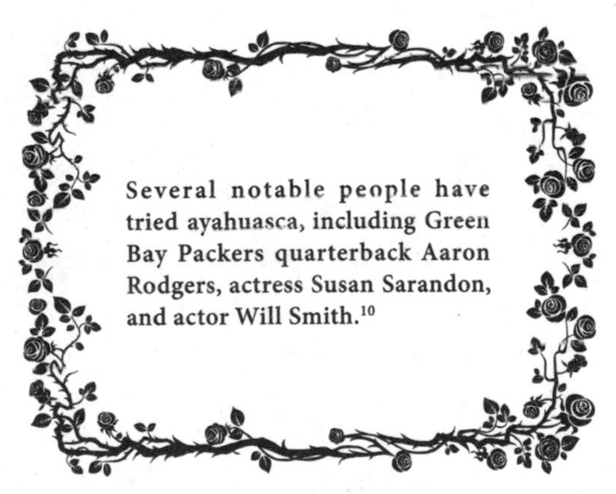

Several notable people have tried ayahuasca, including Green Bay Packers quarterback Aaron Rodgers, actress Susan Sarandon, and actor Will Smith.[10]

The concept of coincidence is brought up when Gwenda meets her uncle in *Sleeping Murder*. How is it explained? According to psychiatrist and coincidence researcher Bernard Beitman, humans may "transmit some unobserved energetic information, which other people then process or organize into emotion and behavior."[11] This theory of synchronicity has existed in scientific studies by psychiatrists and scientists throughout the 1900s, including Paul Kammerer, Albert Einstein, and Carl Jung. Our brains naturally try to create structure out of the things we are experiencing. We start to pick up patterns or similarities in occurrences and start to experience the feeling of coincidence. This becomes even more prominent if we are using selective attention, only paying attention to things that we feel are most relevant or important to us. For example, if you think someone is mad at you, you may start to notice their tone, lack of eye contact, or curt text message in a new light. It may be true, or it may just be a coincidence.

Some of the most famous coincidences in history are truly incredible. First, Violet Jessup happened to be aboard three separate ship disasters, including the sinking of the *Titanic* in 1912, and survived all three! Despite those occurrences, Jessup went on to sail again and died of natural causes in 1971.[12] A second notable coincidence in history involves another one of our favorite authors, Edgar Allan Poe. He published a book in 1838 entitled *The Narrative of Arthur Gordon Pym of Nantucket*. In it, the four-person crew on a ship resort to cannibalism in order to survive. Their victim was a crew boy named Richard Parker. In 1884, a real-life shipwrecked crew, also made up of four people, ate their cabin boy for survival. His name? Richard Parker![13] Last, a set of twins, who were adopted separately, were both named "James" by their respective adoptive parents, both married twice to women named "Linda" and "Betty," and both had

"Edgar Allan Poe created a new literary genre when he wrote 'The Murders in the Rue Morgue.' Although mysteries were not a new literary form, Poe was the first to introduce a character that solved the mystery by analyzing the facts of the case."[14]

sons they named "James Allan" and "James Alan." The coincidences don't end there—they each grew up with adopted brothers named "Larry" and they both had dogs named "Toy."[15]

Miss Marple is a natural conversationalist in Christie's stories and people seem to open up to her quite willingly most of the time. Is this an inherent talent or can we all learn to be better in conversations? According to NPR host Terry Gross, it's important to be genuinely curious about the person you are conversing with. If you want to know the answers to the questions you are asking, it goes a long way. Next, make sure that the conversation is fun! It's good to have a sense of humor and keep the conversation light. Third, organize your thoughts whether you are asking the questions or being interviewed. Last, Gross recommends shifting the conversation to things you want to talk about, if or when appropriate, so that you feel most comfortable.[16] Safe topics for small talk include the weather, sporting events, and other noncontroversial themes.

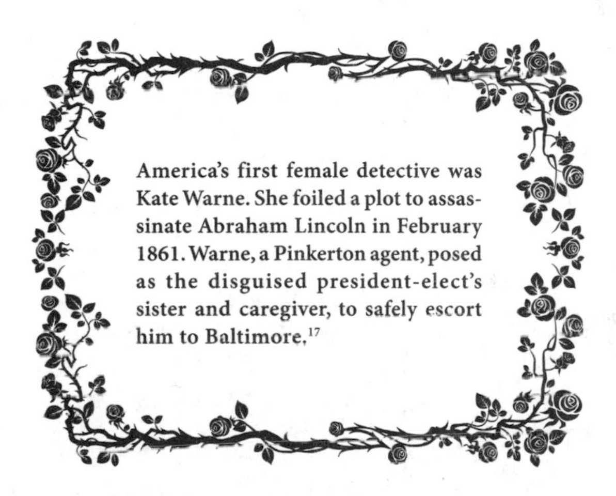

America's first female detective was Kate Warne. She foiled a plot to assassinate Abraham Lincoln in February 1861. Warne, a Pinkerton agent, posed as the disguised president-elect's sister and caregiver, to safely escort him to Baltimore.[17]

In this final book that Miss Marple appears in, unlike her fictional counterpart, Hercule Poirot, she doesn't die in the end. The representation of a sharp, active woman over the age of sixty in a book and on screen is still rare. According to the Geena Davis Institute:

> Analysis of the most popular films and television shows from 2010 to 2020 suggests that on-screen ageism persists and is particularly

> evident among on-screen women aged fifty plus. Just one in four characters who are fifty plus are women, a serious inequality in the representation of older adults in film and television. Moreover, fifty plus women who are on-screen are commonly cast in supporting and minor roles and are less likely to be developed as characters in interesting ways.[18]

The study also found that consumers want more representation of older individuals on screen including those with disabilities, those in romantic relationships, LGBTQIA+ individuals, and older people portrayed as heroes, not villains. Agatha Christie certainly provides an older hero in Miss Marple, and we hope to see more characters like her in the future!

Conclusion

This book has come to an end, yet Agatha Christie's impressive legacy shines on. Like her beloved Egyptian pyramids, Christie's work endures the passage of time, somehow remaining relevant no matter how much society shifts. Perhaps it is because her stories transcend their settings, amplifying what makes us human. Hercule Poirot is egotistical, yet brilliant. Miss Marple is observant, though elderly. These characters are well drawn, and the murders they find themselves party to are even more expertly curated. Most of us haven't lived through a poisoning attempt at a lavish mansion on the Bristol coast, but Agatha Christie's fiction works its way into nearly every home, every library. As the author said herself, there is a reason there is a pervasiveness to her work: "Every murderer is probably someone's old friend."[1] Just remember that the next time you're invited to a dinner party and all the lights go out.

Acknowledgments

Thank you to Nicole and everyone at Skyhorse!

Thank you to our families and friends for coming out to support us in person, virtually, and every way in between!

We are so thankful for Stacey, Karmen, and Mickey for helping us along the way and always looking out for our best interests!

And to our Rewinders, we'll see you in the horror section!

About the Authors

Kelly Florence teaches communication at Lake Superior College in Duluth, Minnesota, and is the creator of the *Be a Better Communicator* podcast. She received her BA in theatre from the University of Minnesota-Duluth and her MA in communicating arts from the University of Wisconsin-Superior. She has written, directed, produced, choreographed, and stage managed for dozens of productions in Minnesota including *Carrie: The Musical* through Rubber Chicken Theatre and *Treasure Island* for Wise Fool Theater. She is passionate about female representation in all media and particularly the horror genre.

Horror and suspense author **Meg Hafdahl** is the creator of numerous stories and books. Her fiction has appeared in anthologies such as *Eve's Requiem: Tales of Women, Mystery, and Horror* and *Eclectically Criminal*. Her work has been produced for audio by *The Wicked Library* and *The Lift*, and she is the author of three popular short story collections including *Twisted Reveries: Thirteen Tales of the Macabre*. Meg is also the author of the novels *This World is Nothing But Evil*, *The Darkest Hunger*, *Daughters of Darkness*, and *Her Dark Inheritance*, called "an intricate tale of betrayal, murder, and small-town intrigue" by *Horror Addicts* and "every bit as page turning as any King novel" by *RW* magazine. Meg lives in the snowy bluffs of Minnesota.

Together, Kelly and Meg have written six books: *The Science of Monsters*, *The Science of Women in Horror*, *The Science of Stephen King*, *The Science of Serial Killers*, *The Science of Witchcraft*, and *The Science of Agatha Christie*. They co-host the *Horror Rewind* podcast and write and produce horror projects together.

Endnotes

Chapter One: The Mysterious Affair at Styles

1. Christie, Agatha. (1977) *Agatha Christie: An Autobiography.* Dodd, Mead, and Company.
2. Ibid.
3. Christie, Agatha. (1920) *The Mysterious Affair at Styles.* John Lane.
4. Ibid.
5. (April 13, 2016) "Interview: David Suchet." *The Strand Magazine.*
6. (August 27, 2018) "How We Judge Personality from Faces Depends on Our Pre-Existing Beliefs About How Personality Works." *NYU.edu.*
7. Crocq, Marc-Antoine MD. (March 2000) "From shell shock and war neurosis to posttraumatic stress disorder: a history of psychotraumatology." *National Library of Medicine.*
8. (2022) "Post-traumatic Stress Disorder." *Mayo Clinic.*
9. Mendelson, Wallace B. MD. (August 16, 2018) "Understanding Sleeping Pills." *Sleep Review.*
10. (2022) "Prescription Sleeping Pills: What's Right For You?" *Mayo Clinic.org.*
11. (2014) "Strychnine: Last of the Romantic Poisons." *Nature's Poisons.*
12. Phillips, Graham. (2004) *Alexander the Great. Murder in Babylon*. Virgin Books.
13. Kridel, Kristen. (February 12, 2008) "A century-old mystery: Did serial killer fake her death?" *Chicago Tribune.*
14. (2022) "The Mysterious Affair at Styles." *Agatha Christie.com.*
15. Omachi, Fukiko et al. (December 3, 2019) "Relationship between the effects of food on the pharmacokinetics of oral antineoplastic drugs and their physicochemical properties." *Journal of Pharmaceutical Health Care and Sciences.*
16. Goupil, Louise et al. (September 27, 2021) "Listeners' perceptions of the certainty and honesty of a speaker are associated with a common prosodic signature." *Nat Commun.*
17. Olson, Kent R. (November 1, 2003) *Poisoning & Drug Overdose (4th ed.).* Appleton & Lange.
18. (2022) "The Mysterious Affair at Styles." *Agatha Christie.com.*

Chapter Two: The Murder of Roger Ackroyd

1. Hazelton, Pamela. (December 14, 2020) "If Your Writing Isn't Getting Rejected, You're Doing It Wrong." *Medium.com.*
2. Christie, Agatha. (1926) *The Murder of Roger Ackroyd.* Dodd, Mead, and Company.
3. Ibid.
4. Ibid.
5. (2022) "The Unreliable Narrator." *Americanliterature.com.*
6. Geberth, Vernon J. (2013) "The Seven Major Mistakes in Suicide Investigation." *Practical Homicide.*
7. (2008) "Historical Leading Causes of Death." *National Center for Health Statistics.*
8. (2022) "Gastritis." *Mayo Clinic.*
9. (2022) "History of Keys—Who Invented Keys?" *History of Keys.com.*
10. (2021) "Footprints." *Crime Museum.*
11. (July 19, 2014) "It's a Fact—Our Feet Are Getting Bigger." *Walk EZ Store.*

Chapter Three: The Seven Dials Mystery

1. (December 5,1926) "Mrs. Agatha Christie, Novelist, Disappears in Strange Way From her Home in England." *The New York Times.*
2. (December 9, 1926) "500 Police and Planes Hunt for Mrs. Christie." *The New York Times.*
3. Wilson, Andrew. (May 6, 2017) "Solved, the last great Agatha Christie mystery? Acclaimed biographer presents chilling new theory about crime novelist's disappearance." *Daily Mail.*
4. (2022) "Wellness Industry Statistics & Facts." *Global Wellness Institute.org.*
5. Christie, Agatha. (1929) *The Seven Dials Mystery.* Dodd, Mead, and Company.
6. Parker, Hilary. (January 15, 2022) "Physical Side Effects of Oversleeping." *Web MD.com.*
7. Rosseland, Ragna et al. (October 31, 2018) "Effects of Sleep Fragmentation and Induced Mood on Pain Tolerance and Pain Sensitivity in Young Healthy Adults." *Frontiers in Psychology.*
8. Russo, Naomi. (April 21, 2006) "A 2,000-Year History of Alarm Clocks." *Atlas Obscura.*
9. Bryant, Clifton D. (2003) *Handbook of Death & Dying.* Thousand Oaks, CA: Sage Publications.
10. Weaver, Matthew. (September 9, 2019) "Cloud Atlas Child Actor Died by Misadventure, Coroner Rules." *The Guardian.*
11. (2022) "Fire and Arson Investigations." *National Institute of Justice.*
12. Mohapatra, Mounabati et al. (2017) "Bitemarks in Forensic Odontology." *Indian Journal of Forensic Odontology.*
13. (2022) "Early Forensic Odontology." *National Museum of Dentistry.*

14. (2022) "Bitemark Evidence." *California Innocence Project.*
15. Shachtman, Noah. (November 16, 2012) "They Cracked This 250-Year-Old Code, and Found a Secret Society Inside." *Wired.*
16. Mansky, Jackie. (March 7, 2016) "Eight Secret Societies You Might Not Know." *Smithsonian Magazine.*

Chapter Four: The Sittaford Mystery

1. Christie, Agatha. (1931) *The Sittaford Mystery.* Collins Crime Club.
2. (2022) "Ur." *Britannica.com.*
3. (2022) "The Lure of the Red Herring." *Worldwidewords.org.*
4. (October 27, 2021) "Number of Escapees From Prisons in the U.S. 2000–2019." *Statista.*
5. Miller, Claire. (August 13, 2019) "This is How Many Criminals Have Escaped From Prison in Gloucestershire in the Last Year—and How Many Have Been Caught." *Gloucestershire Live.*
6. (2022) "Did You Know?" *Casino.org.*
7. (2022) "History of Corrective Lenses." *Glasses History.*
8. (June 14, 2016) "The Science of Eyeglasses." *Williams Eye Works.*
9. (2022) "What Percentage of the Population Wears Glasses?" *Glasses Crafter .com.*
10. Rodriguez McRobbie, Linda. (October 27, 2013) "The Strange and Mysterious History of the Ouija Board." *Smithsonian Magazine.*
11. (2022) "What To Do If You're Caught in a Winter Storm." *Weather.gov.*
12. (November 5, 2020) "The Disturbing Truth Behind a Sardonic Grin." *Ancient Origins.*
13. Janos, Adam. (February 11, 2021) "How is Time of Death Determined for a Crime Scene Victim? Hint: It Usually Involves Bugs." *A&E TV.com.*
14. Laney, Cara and Loftus, Elizabeth F. (2022) "Eyewitness Testimony and Memory Biases." *NOBA Project.*
15. Stromberg, Joseph. (May 17, 2012) "The Science of Sleepwalking." *Smithsonian Magazine.*

Chapter Five: Murder on the Orient Express

1. Wagstaff, Vanessa. (2004) *Agatha Christie: A Reader's Companion* Aurum.
2. Zax, David. (November 7, 2017) "The True History of the Orient Express." *Smithsonian Magazine.*
3. Christie, Agatha. (1977) *Agatha Christie: An Autobiography.* Dodd, Mead and Company.
4. Ibid.
5. Christie, Agatha. (1934) *Murder on the Orient Express.* Collins Crime Club.
6. Pariona, Amber. (April 25, 2017) "Fluvial Landforms: What is a Wadi?" *WorldAtlas.com.*
7. (2022) "Why Do My Chicken Eggs Look Different?" *Centre Hall Farm Store.com.*

8. (November 10, 2018) "Kenneth Branagh On His Meticulous Master Detective Role In 'Murder On The *Orient Express*.'" *NPR*.
9. Barton, Laura. (May 18, 2009) "Poirot and Me." *The Guardian*.
10. (2022) "Obsessive Compulsive Disorder." *Mayo Clinic*.
11. Dempsey, Caitlin. (January 29, 2011) "Mapping Through the Ages: The History of Cartography." *GIS Lounge*.
12. Ibid.
13. Evans, Holly. (April 16, 2022) "Brutal Murder of Woman, 26, On south London train remains unsolved 34 years later despite killer's blood left at scene." *Mylondon.news*.
14. O'Brian, Bridget. (2019) "The Long and Strange History of Celebrity." *Columbia Magazine*.
15. Blumhardt, Miles. (January 21, 2022) "White Death: Avalanche Deaths Persist in Colorado, Across US Despite Improved Forecasting." *Coloradoan*.
16. (1970) "The Peru Earthquake: A Special Study." *Bulletin of the Atomic Scientists*.
17. Howard, Jenny. (July 19, 2019) "Avalanches, Explained." *National Geographic*.
18. Hardey, Sarah. (October 26, 2021) "Barbiturate Overdose: Symptoms, Effects, and Risks." *American Addiction Centers*.
19. (2022) "How to Help Someone Who's Been Stabbed." *Red Cross.org*.
20. (July 27, 2018) "Grief: What's Normal, What's Not—and 13 Tips to Get Through It." *Cleveland Clinic*.

Chapter Six: The Murder at the Vicarage

1. Christie, Agatha. (1930) *The Murder at the Vicarage*. Collins Crime Club.
2. Ibid.
3. Ibid.
4. Bolin, Alice. (May 15, 2015) "Ms. Marple vs. The Mansplainers: Agatha Christie's Feminist Detective Hero." *Electricliterature.com*.
5. Gill, Gillian. (1990) *Agatha Christie: The Woman and Her Mysteries*. The Free Press.
6. Ibid.
7. (January 10, 2020) "How Joan Hickson became the ultimate Miss Marple." *Vision.tv.ca*.
8. Ibid.
9. Bowen, Rhys. (September 1, 2022) "Miss Marple is Agatha Christie's best character. A new book reminds us why." *The Washington Post*.
10. Christie, Agatha. (1977) *Agatha Christie: An Autobiography*. Dodd, Mead, and Company.
11. Gulla, Emily. (February 14, 2020) "A Brief History of the word 'Spinster' and how it's used today." *Cosmopolitan*.
12. Hall, Edward T. (1966) *The Hidden Dimension*. Anchor Books.

13. Chamberlin, Jamie. (2008) "The Time of Our Lives." *American Psychological Association.*
14. Bolin, Alice. (May 15, 2015) "Ms. Marple vs. The Mansplainers: Agatha Christie's Feminist Detective Hero." *Electricliterature.com.*
15. Oliveira LS, Justino E, Freitas C, et al. (2005) "The Graphology Applied to Signature Verification."
16. DeKalb Miller, Meredith. (2022) "Forensic Handwriting and Signature Analysis." *Dekalb Miller.com.*
17. (2022) "What Does Your Handwriting Say About You?" *Pens.com.*

Chapter Seven: Death on the Nile

1. Wong, Jun Yi. (November 9, 2021) "Agatha Christie Dug for Clues for Real in Egypt." *New Lines Magazine.*
2. Morris, David. (April 23, 2022) "Insights: Christie's still unpublished works." *Collectingchristie.com.*
3. Solly, Meilan. (February 10, 2022) "How Agatha Christie's love of archeology influenced 'Death on the Nile.'" *Smithsonian Magazine.*
4. Wong, Jun Yi. (November 9, 2021) "Agatha Christie Dug for Clues for Real in Egypt." *New Lines Magazine.*
5. (September 22, 2022) *Steam-ship-sudan.com.*
6. Christie, Agatha. (1937) *Death on the Nile.* Dodd, Mead and Company.
7. Solly, Meilan. (February 10, 2022) "How Agatha Christie's love of archeology influenced 'Death on the Nile.'" *Smithsonian Magazine.*
8. Daley, Beth. (May 22, 2018) "Agatha Christie: World's first historical whodunnit was inspired by 4,000 year old letters." *The Conversation.*
9. (September 22, 2022) *Agathachristie.com.*
10. (September 22, 2022) "Heqanakhte Letter I." *Metmuseum.com.*
11. (April 11, 2022) "Weather Proverbs About Birds." *The Farmer's Almanac.*
12. Willems, Aimee. (2022) "Fowl Forecast—How Birds Predict Weather." *More Birds.com.*
13. Yanes, Javier. (January 29, 2018) "Physics Unveils the Mysteries of the Egyptian Pyramids." *Open Mind.*
14. Stewart-Williams, Steve. (December 6, 2018) "Where Does Jealousy Come From?" *Science Focus.*
15. Hanlon, Sarah. (March 3, 2022) "This is the Average Cost of a Honeymoon Today." *The Knot.*
16. Johnson, Samuel. (1818) *A Dictionary of the English Language.* Longman, Hurst, Rees, Orme, and Brown.
17. Levenson, Rachel. (March 17, 2021) "How Long Does the Honeymoon Phase Last, According to 4 Experts." *Up Journey.*
18. Cunningham, Ed. (June 14, 2022) "These are officially the most popular honeymoon destinations for 2022." *Timeout.com.*

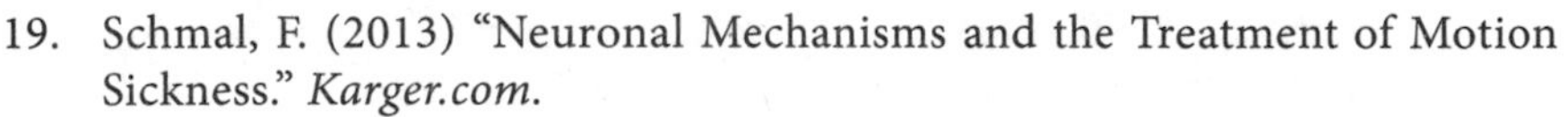

19. Schmal, F. (2013) "Neuronal Mechanisms and the Treatment of Motion Sickness." *Karger.com.*
20. Erskine, Stefanie K. (2022) "Motion Sickness." *CDC.gov.*
21. Shrestha R, Kanchan T, Krishan K. (May 15, 2022) *Gunshot Wounds Forensic Pathology.* StatPearls Publishing.
22. (2022) "Bloodstain Pattern Analysis." *Forensic Science Simplified.org.*
23. Ibid.
24. Bates, Brandon. (April 2, 2019) "Expert: 'The average person doesn't realize how far a bullet from a gun travels.'" *WBIR News.*

Chapter Eight: And Then There Were None

1. Christie, Agatha. (1939) *And Then There Were None.* Dodd, Mead, and Company.
2. Grabinowski, Ed. (September 26, 2022) "The 21 Best Selling Books of all Time." *Howstuffworks.com.*
3. Gill, Gillian. (1990) *Agatha Christie, The Woman and her Mysteries.* The Free Press.
4. (September 26, 2022) "Serial Killers vs. Mass Murderers." *Crimemuseum.org.*
5. Christie, Agatha. (1939) *And Then There Were None.*
6. Christie, Agatha. (1977) *Agatha Christie: An Autobiography.* Dodd, Mead, and Company.
7. (September 26, 2022) "Minstrel Songs." *The Library of Congress.*
8. Elflein, John. (February 1, 2022) "Number of Choking-Deaths in the U.S. 1945–2020." *Statista.com.*
9. (2022) "Choking Prevention and Rescue Tips." *The National Safety Council.*
10. Olson, Ted. (2004) *Crossroads: A Southern Culture Annual.* Mercer University Press.
11. Banner, Lois. (2012) *Marilyn: The Passion and the Paradox.* Bloomsbury.
12. Sedensky, Matt. (March 26, 2007) "Smith Died From Accidental Drug Overdose." *Associated Press.*
13. Quealy, Kevin. (December 5, 2015) "In Other Countries, You're as Likely to Be Killed by a Falling Object as by a Gun." *The New York Times.*
14. (Aug 9, 2022) "Aug 22nd, 564 CE Loch Ness Monster Sighted." *Nationalgeographic.com.*
15. Herz, Rachel S. (November 11, 2002) "Do Scents Affect People's Moods or Work Performance?" *Scientific American.*
16. Walsh, Colleen. (February 27, 2020) "What the Nose Knows." *The Harvard Gazette.*
17. Shockman, Elizabeht and Minoff, Annie. (October 10, 2015) "Agatha Christie's Murders are Enmeshed With Real Chemistry." *The World.org.*

Chapter Nine: Murder in Retrospect

1. Christie, Agatha. (1942) *Five Little Pigs.*

2. Gill, Gillian. (1990) *Agatha Christie: The Woman and Her Mysteries*. The Free Press.
3. Christie, Agatha. (1934) *Unfinished Portrait.*
4. Morgan, Janet. (November 19, 2004) "Rosalind Hicks." *TheGuardian.com.*
5. Christie, Agatha. (1977) *Agatha Christie: An Autobiography.* Dodd, Mead, and Company.
6. Christie, Agatha. (1942) *Five Little Pigs.*
7. (October 6, 2022) "What is a Cold Case?" *Houstontx.gov.*
8. Christie, Agatha. (1942) *Five Little Pigs.*
9. Gross, Samuel R. et al. (September 1, 2020) "Government Misconduct and Convicting the Innocent." *National Registry of Exonerations.*
10. Subramanian, Ram. (November 29, 2021) "How Some European Prisons Are Based on Dignity Instead of Dehumanization." *Brennan Center.org.*
11. Pilkington, Ed. (April 28, 2022) "A Bite Mark, a Forensic Dentist, a Murder: How Junk Science Ruins Innocent Lives." *The Guardian.*
12. R. G. Frey. (1978) "Did Socrates Commit Suicide?" *Philosophy.*
13. Grieve, M. (1971) *A Modern Herbal (2nd ed.).* Mineola, NY: Dover Publications.
14. Mullen, Paul E. (1995) "Jealousy and Violence." *Hong Kong Journal of Psychiatry.*
15. Ramachandran VS, Jalal B. (September 1, 2017) "The Evolutionary Psychology of Envy and Jealousy." *Front Psychol.*

Chapter Ten: The Crooked House

1. "There was a Crooked Man." (October 11, 2022) *allnurseryrhymes.com.*
2. Christie, Agatha. (1949) *Crooked House.* Dodd, Mead, and Company.
3. Ibid.
4. Ibid.
5. (July 28, 2011) "Who Mops the Floor Now? How Domestic Service Shaped 20th Century Britain." University of Cambridge. *Cam.ac.uk.*
6. Gill, Gillian. (1990) *Agatha Christie: The Woman and her Mysteries.* The Free Press.
7. Christie, Agatha. (1949) *Crooked House.*
8. (July 28, 2011) "Who Mops the Floor Now? How Domestic Service Shaped 20th Century Britain." University of Cambridge. *Cam.ac.uk.*
9. Christie, Agatha. (1949) *Crooked House.*
10. Craig, Robert. (December 20, 2018) "A History of Syringes and Needles." *The University of Queensland.*
11. (December 5, 2002) "The Irish Doctor Who Invented the Syringe." *The Irish Times.*
12. Beers, Dean A. et al. (2012) *Professional Investigations: Ethical Considerations for the Professional Investigator.* Smash Words.com.
13. Ramrarine, Mityanand MD. (June 6, 2020) "What are common side effects of physostigmine salicylate for anticholinergic toxicity?" *Medscape.*

14. Coelho F, Birks J. (2001) "Physostigmine for Alzheimer's disease." *The Cochrane Database of Systematic Reviews.*
15. (2022) "How Is Alzheimer's Disease Treated?" *National Institute on Aging.*
16. (November 26, 2020) "How to Get Rid of Moles in Your Yard?" *V Extermination.*
17. (February 14, 2017) "'Hair of the Dog' Won't Cure That Hangover." *Science Daily.*
18. (2002) "Early Child Care and Children's Development Prior to School Entry: Results from the NICHD Study of Early Child Care." *American Educational Research Journal.*
19. Im Y, Vanderweele TJ. (2018) "Role of First-Year Maternal Employment and Paternal Involvement in Behavioral and Cognitive Development of Young Children." *Infant Ment Health.*
20. (2021) "Childcare: What the Science Says." *Critical Science, Medium.*
21. Feldman, Justin. (January 22, 2015) "Daycare and Early Childhood Education in the United States: Research Roundup." *The Journalist's Resource.*
22. Cralk, Fergus and Lockhart, Robert. (December 1972) "Levels of Processing: A Framework For Memory Research." *Journal of Verbal Learning and Verbal Behavior.*
23. Dasgupta, Shreya. (May 12, 2016) "How Many Plant Species are There in the World? Scientists Now Have an Answer." *Mongabay News.*
24. (2022) "Becoming a Toxicologist." *Society of Toxicology.*
25. Wolff S, Smith AM. (2001) "Children Who Kill." *BMJ.*

Chapter Eleven: A Murder Is Announced

1. Christie, Agatha. (1950) *A Murder Is Announced.* Dodd, Mead, and Company.
2. (October 26, 2022) "The Jury Box." *NobleKnight.com.*
3. Popovici, Alice. (August 29, 2018) "The Game of Clue was Borne out of Boredom during WWII Air-Raid Blackouts." *History.com.*
4. Christie, Agatha. (1950) *A Murder Is Announced.*
5. Ibid.
6. Ibid.
7. Lahiri, Tripti. (March 25, 2019) "Why it's a crime to download or print the mosque shooter's manifesto in New Zealand." *Quartz.*
8. Arango, Tim et al. (August 3, 2019) "Minutes Before El Paso Killing, Hate-Filled Manifesto Appears Online." *The New York Times.*
9. Collins, Ben. (May 14, 2022) "The Buffalo supermarket shooting suspect allegedly posted an apparent manifesto repeatedly citing 'great replacement' theory." *NBC News.*
10. (2022) "What You Need To Know About Rationing In The Second World War." *Imperial War Museums.*

11. Coats, Kenneth. (August 14, 2018) "The Future of Policing Using Pre-Crime Technology." *Forbes.*
12. (2022) "Minority Report." *Wikipedia.org.*
13. Watercutter, Angela. (December 7, 2018) "How the CIA Trains Spies to Hide in Plain Sight." *Wired.*
14. Poddar, Aanchai. (June 29, 2022) "Why You Shouldn't Ignore Face Blindness that Brad Pitt & Shenaz Treasury Suffer From." *The Indian Express.*
15. (March 6, 2021) "Foods Linked to Better Brainpower." *Harvard Medical School.*
16. Ladd, Kara. (September 25, 2018) "Is Chromotherapy the Real Deal?" *Architectural Digest.*
17. (2022) "The Science of Memory." *John Hopkins Medicine.*
18. Fitch, Dr. Brian. (January 6, 2015) "Memory Puzzle: What Every Investigator Should Know." *LEB FBI.gov.*
19. Null, Christopher. (August 1, 2006) "The Best: Deadly Poison, Ingested or Inhaled." *Wired.*
20. Garner, Ross. (2022) "What Happens When Electrical Wiring Gets Wet?" *Hunker.*
21. (2022) "Thyroidectomy." *Mayo Clinic.org.*
22. Huan, Song et al. (2020) "Risk of Psychiatric Disorders Among the Surviving Twins After a Co-Twin Loss." *eLife.*
23. Irvine, Martha. (July 17, 1997) "Sister Accused of Killing Sibling, Assuming Her Identity." *AP News.*

Chapter Twelve: They Do It with Mirrors

1. Christie, Agatha. (1952) *They Do It with Mirrors*. Dodd, Mead, and Company.
2. Ibid.
3. Christie, Agatha. (1950) *A Murder Is Announced*. Dodd, Mead, and Company.
4. Christie, Agatha. (1952) *They Do It with Mirrors*. Dodd, Mead, and Company.
5. Gill, Gillian. (1990) *Agatha Christie: The Woman and Her Mysteries*. The Free Press.
6. Staveley-Wadham, Rose. (August 28, 2019) "Requisitioning of Country Houses in the Second World War—Hospitals, War Supply Depots and More." *The British Newspaper Archive.*
7. Bradley, Kate. (October 2008) "Juvenile delinquency and the evolution of the British juvenile courts, c.1900–1950." *Archives.History.AC.UK.*
8. Christie, Agatha. (1952) *They Do It with Mirrors*. Dodd, Mead, and Company.
9. Landers, Jackson. (September 27, 2016) "In the Early 19th Century, Firefighters Fought Fires . . . and Each Other." *Smithsonian Magazine.*

10. Covington, Taylor. (May 18, 2022) "House Fire Statistics and Facts 2022." *The Zebra.*
11. Rossen, Jeff. (January 14, 2016) "Newer Homes and Furniture Burn Faster, Giving You Less Time to Escape a Fire." *Today.com.*
12. Orenstein, Beth W. (July 21, 2020) "3-Step Fitness Plan for Psoriatic Arthritis." *Everyday Health.*
13. Peake, Jonathan W. et al. (November 13, 2016) "The Effects of Cold Water Immersion and Active Recovery on Inflammation and Cell Stress Responses in Human Skeletal Muscle After Resistance Exercise." *The Journal of Physiology.*
14. Wells, Rachel. (2022) "5 Ways to Stop Panic in Its Tracks." *Happify.com.*
15. Horton, D.; Wohl, R. (1956) "Mass Communication and Para-Social Interaction: Observation on Intimacy at a Distance." *Psychiatry.*
16. Brooks, Samantha K. (September 26, 2018) "FANatics: Systematic Literature Review of Factors Associated With Celebrity Worship, and Suggested Directions for Future Research." *Current Psychology.*
17. Kapelovitz, Leonard H. (1987) *To Love and To Work: A Demonstration and Discussion of Psychotherapy.* Jason Aronson, Inc.
18. Debrowski, Adam. (2022) "How Does Night Vision Work?" *All About Vision.*
19. (February 21, 2021) "A Science Teacher Explains: How Do Some Animals See Better at Night?" *The Indian Express.*
20. Ratnaike, R. N. (July 1, 2003) "Acute and Chronic Arsenic Toxicity." *Postgraduate Medical Journal.*
21. (2022) "Fast Facts About Food Poisoning." *CDC.gov.*
22. (2022) "Oyster Facts." *Destination Panama City.*
23. Badiye, Ashish, et al. (November 14, 2021) "Forensic Gait Analysis." *National Library of Medicine.*
24. Adams, Cecil. (April 13, 2001) "Do Crime Scene Investigators Really Draw a Chalk Line Around the Body?" *The Straight Dope.*
25. (March 2016) "Recognizing Magic as a Rare and Valuable Art Form and National Treasure." *Congress.gov.*

Chapter Thirteen: The Mousetrap

1. (November 2, 2022) "The Mousetrap." *AgathaChristie.com.*
2. Bugbee, Teo. (September 15, 2022) "'See How They Run' Review: An Agatha Christie Mystery Spoof" *The New York Times.*
3. Christie, Agatha. (1979) *Agatha Christie: An Autobiography.* Dodd, Mead, and Company.
4. Ibid.
5. Shakespeare, William. *Hamlet.*
6. Rowe, Michael. et al. (2018) "A Pilot Study of Motive Control to Reduce Vengeance Cravings." *The Journal of the American Academy of Psychiatry and the Law.*

7. Rowe, Michael. et al. (2018) "A Pilot Study of Motive Control to Reduce Vengeance Cravings." *The Journal of the American Academy of Psychiatry and the Law*.
8. Marguerite, Tassi. (September 22, 2012) "Women and Revenge in Shakespeare: Gender, Genre, and Ethics." *Renaissance Quarterly*.
9. (2022) "Dialect Coach." *Careers Broadway*.
10. Miller, Wilbur. (2012) *The Social History of Crime and Punishment in America: A-De*. Thousand Oaks, CA: Sage.

Chapter Fourteen: The Pale Horse

1. "Ariadne Oliver." (November 3, 2022) *AgathaChristie.com*.
2. Gill, Gillian. (1990) *Agatha Christie: The Woman and her Mysteries*. The Free Press.
3. Blakemore, Erin. (September 15, 2016) "Agatha Christie: Pharmacist." *JSTOR Daily*.
4. Twilley, Nicola. (September 8, 2015) "Agatha Christie and the Golden Age of Poisons." *The New Yorker*.
5. Christie, Agatha. (1977) *Agatha Christie: An Autobiography*. Dodd, Mead, and Company.
6. Ibid.
7. Nutt, Amy Ellis et al. (February 13, 2011) "A 15-year-old Case Yields a Timely Clue in Deadly Thallium Poisoning." *The Star Ledger*.
8. Sanders, Dennis; Lovallo, Len. (1984) *The Agatha Christie Companion*. Delacorte.
9. Emsley, John. (April 28, 2005) *The Elements of Murder: A History of Poison*. OUP Oxford.
10. Mankey, Jason. (August 5, 2014) "Pagan Time Capsule: 1950's." *Patheos.com*.
11. Palmer, Ewan. (February 15, 2022) "Pastor Greg Locke Threatens to Expose 'Witches' in His Church in Viral Video." *Newsweek*.
12. Stych, Anne. (March 4, 2022) "Greg Locke, Pastor Who Called out Witches During Sunday Meeting, Says He's been Bombarded with Profane Items." *Ministry Watch*.
13. (2022) "Benjamin Abbot House." *Salem Witch Museum.com*.
14. (2022) "Salem Witch Trials Memorial." *Salem Witch Museum.com*.
15. (2022) "The Witch Monument in Vardø is in Memory of the 91 Witch Trial Victims." *Nord norge.com*.
16. Brooks, Libby. (October 29, 2019) "Calls for Memorial to Scotland's Tortured and Executed Witches." *The Guardian*.
17. Campsie, Alison. (March 24, 2022) "Campaigners Find 'Ideal Location' for National Memorial to Scotland's 'Witches'." *The Scotsman*.
18. Zamula, E. (1991) "A New Challenge for Former Polio Patients". F*DA Consumer*.
19. (2008) "Malingering." *Gale Encyclopedia of Medicine*.

20. Avila, Jim and Cohen, Deirdre. (July 22, 2010) "Medical Mystery or Hoax: Did Cheerleader Fake a Muscle Disorder?" *ABC News*.
21. Ibid.
22. Llera, Ryan. (2022) "Ringworm in Dogs." *VCA Animal Hospitals*.
23. (2022) "Treatment for Ringworm." *CDC.gov*.
24. Mukhopadhyay, Amiya Kumar. (November 5, 2018) "A Historical Note on the Evolution of 'Ringworm.'" *Indian Journal of Dermatology, Venereology, and Leprology*.
25. (2022) "Hair Loss." *Mayo Clinic*.
26. (2022) "Hair loss: Diagnosis and Treatment." *American Academy of Dermatology.org*.

Chapter Fifteen: The Clocks

1. Gill, Gillian. (1990) *Agatha Christie: The Woman and her Mysteries*. The Free Press.
2. Christie, Agatha. (1977) *Agatha Christie: An Autobiography*. Dodd, Mead, and Company.
3. (November 7, 2022) "Poirot's Age." *Poirot.us*.
4. Christie, Agatha. (1963) *The Clocks*. Dodd, Mead, and Company.
5. Christie, Agatha. (1977) *Agatha Christie: An Autobiography*. Dodd, Mead, and Company.
6. Christie, Agatha. (1937) *Dumb Witness*. Dodd, Mead, and Company.
7. Prichard, Mathew. (November 7, 2022) "The Essence of Agatha Christie: Dogs" *Agathachristie.com*.
8. (November 7, 2022) "Christie's Critters: Assistants, Avengers, and Animal Imagery." *Crossexaminingcrime.Wordpress.com*.
9. Christie, Agatha. (1977) *Agatha Christie: An Autobiography*. Dodd, Mead, and Company.
10. (2022) "Apprenticeship to a Shorthand—Writer." *Papyri.info*.
11. Bellis, Mary. (February 6, 2019) "The Development of Clocks and Watches Over Time." *Thought Co.com*.
12. Walsh, Michael. (October 28, 2020) "Learn the Many Ways Cultures Around the World Measure Time." *Nerdist*.
13. (August 13, 2018) "My Antique Clock Runs Slow—Why?" *Antique and Vintage Clock.com*.
14. (2022) "History of Braille." *Braille Works.com*.
15. (May 24, 2022) "What are Typical Accommodations for Students with Blindness?" *Washington.edu*.

Chapter Sixteen: By the Pricking of My Thumbs

1. Shakespeare, William. *Macbeth*.
2. Stein, Sadie. (February 4, 2013) "When Agatha Christie was Investigated by MI5." *TheParisReview.org*.

3. Gill, Gillian. (1990) *Agatha Christie: The Woman and Her Mysteries*. The Free Press.
4. Christie, Agatha. (1968) *By the Pricking of My Thumbs*. Dodd, Mead, and Company.
5. Esposito, Veronica. (January 19, 2022) "Mapping Fiction: The Complicated Relationship between Authors and Literary Maps." *TheGuardian.com*.
6. Autencio, Kristel. (September 28, 2015) "Grounds for Murder: Maps and Floor Plans in Mystery Novels." *Bookriot.com*.
7. (2022) "The Science Behind Concussions." *XLNT Brain.com*.
8. (2022) "What is a Concussion?" *Concussion Foundation.org*.
9. (May 31, 2016) "Poll: Nearly 1 In 4 Americans Reports Having Had A Concussion." *NPR News*.
10. (2022) "Why are Escape Rooms so Popular?" *World of Escapes.com*.
11. Williams, Anne D. (2022) "Origins of Jigsaw Puzzles." *Puzzle Warehouse*.
12. Spira, Lisa. (July 30, 2022) "US Escape Room Industry Report—July 2022." *Room Escape Artist.com*.

Chapter Seventeen: Hallowe'en Party

1. Christie, Agatha. (1969) *Hallowe'en Party*. Dodd, Mead, and Company.
2. Barber, Nicholas. (December 23, 2020) "The Man Who Wrote the Most Perfect Sentences Ever Written." *BBC.com*.
3. Hastings, Chris. (July 14, 2018) "'Thanks for all the laughs': Touching 'last letter' from Agatha Christie to 'favourite writer' PG Wodehouse reveals firm friendship they formed in their later years." *The Daily Mail*.
4. (October 8, 2012) "Mystery behind Agatha Christie's missing 'damehood' finally solved" *BigNewsNetwork.com*.
5. Morgan, Joan. (2002) *The New Book of Apples*. Ebury Press.
6. (2022) "Halloween." *History.com*.
7. Soo Hoo, Fawnia. (November 10, 2017) "How the 'Murder on the Orient Express' Costume Designer Outfitted Daisy Ridley and Michelle Pfeiffer in Authentic '30s Clothing." *Fashionista.com*.
8. (October 2008) "Fashion Designing – the Then and Now." *Fibre 2 Fashion .com*.
9. (August 20, 1963) "Lucinda Ballard Obituary." *The New York Times*.

Chapter Eighteen: Elephants Can Remember

1. Christie, Agatha. (1972) *Elephants Can Remember*. Dodd, Mead, and Company.
2. Flood, Allison. (April 3, 2009) "Study Claims Agatha Christie had Alzheimer's." *The Guardian*.
3. Abumrad, Jad. (June 1, 2010) "Agatha Christie and Nuns Tell a Tale of Alzheimer's." *NPR.org*.
4. Ibid.

5. Ibid.
6. Ibid.
7. Perry, Susan (May 6, 2010) "Mystery Solved? Agatha Christie's Later Novels Suggest the Onset of Dementia." Minnpost.com.
8. Smith, Matthew, PhD. (October 9, 2015) "The Healing Waters." *Psychology Today.*
9. (2022) "Water Use in Hydrotherapy Tanks." *CDC.gov.*
10. Huttunen P, Kokko L, Ylijukuri V. (2004) "Winter Swimming Improves General Well-Being." *Int J Circumpolar Health.*
11. (June 17, 2022) "Elephant Kills Woman and Returns to Her Funeral to Attack Her Corpse." *Fox 29 News.*
12. (2022) "Do Elephants Ever Forget?" *Wonderopolis.*
13. (June 30, 2011) "Never Cross a Crow—It Will Remember Your Face." *The Conversation.*
14. Panko, Ben. (November 29, 2016) "Dogs May Possess a Type of Memory Once Considered 'Uniquely Human.'" *Smithsonian Magazine.*
15. Hauser, Dr. Wendy. (2022) "What to Know About Mental Health Care and Dogs." *ASPCA.*
16. Ibid.
17. Chiwaya, Nigel et al. "Interactive: What Solved Murder Data Says About Homicides in the U.S." *NBC News.*
18. (2022) "Dementia Facts and Figures." *Alzheimer's Disease International.org.*

Chapter Nineteen: Curtain

1. Gill, Gillian. (1990) *Agatha Christie: The Woman and Her Mysteries.* The Free Press.
2. Christie, Agatha. (1977) *Agatha Christie: An Autobiography.* Dodd, Mead, and Company.
3. Carr, Flora. (October 7, 2020) "Did Agatha Christie really hate Poirot? Channel 5 drama writer reveals all." *Radiotimes.com.*
4. Jeffery, Morgan. (November 11, 2013) "David Suchet says Goodbye to 'Poirot': The Actor on Making TV History." *Digitalspy.com.*
5. Podesta, James. (March 4, 2022) "Sussex seller auctioning off £50k Agatha Christie treasure trove—including mysterious Valentine's Day card." *Sussex.live.co.uk.*
6. Watt, Jeff. (October, 2008) "Buddhist Deity: Amitayus Buddha." *Himalyianart.com.*
7. (2022) "Arthritis." *CDC.gov.*
8. (2022) "Juvenile Arthritis." *American College of Rheumatology.*
9. Trickey, Erik. (January 4, 2018) "Inside the Story of America's 19th-Century Opiate Addiction." *Smithsonian Magazine.*
10. (2022) "Overdose Death Rates." *National Institute on Drug Abuse.*

11. (2022) "Does Morphine Make Death Come Sooner?" *Canadian Virtual Hospice.*
12. Delaterre, Flora. (May 12, 2014) "Calabar Bean." *Flora Delaterre—Plant Information.*
13. Davis KL, Mohs RC, Tinklenberg JR, Pfefferbaum A, Hollister LE, Kopell BS. (1978) "Physostigmine: improvement of long-term memory processes in normal humans." *Science.*
14. (2022) "'Green Eyed Monster', Meaning & Context." *No Sweat Shakespeare.*
15. (2022) "Curtain: Poirot's Last Case." *Agathachristie.com.*
16. (2022) "Past Poirot Portrayals." *Agatha Christie.com.*

Chapter Twenty: The Sleeping Murder

1. Christie, Agatha. (1977) *Agatha Christie: An Autobiography.* Dodd, Mead, and Company.
2. Ibid.
3. (November 14, 2022) "How Christie Wrote." *AgathaChristie.com.*
4. Riches, Tony. (February, 2014) "Agatha Christie's Writing Habits." *TonyRiches.com.*
5. (January 6, 2018) "Agatha Christie: The Queen of Crime." *Awritersden. Wordpress.com.*
6. Christie, Agatha. (1977) *Agatha Christie: An Autobiography.* Dodd, Mead, and Company.
7. Henry Otgaar, Mark L. Howe & Lawrence Patihis. (2022) "What Science Tells us About False and Repressed Memories." *Memory, vol. 30.*
8. Paul, Marla. (August 17, 2015) "How Traumatic Memories Hide In The Brain, and How To Retrieve Them." *Northwestern Medicine.*
9. Margolin, Madison. (August 31, 2016) "This is How Ayahuasca Affects the Brain." *Vice.com.*
10. White, Jeanae. (November 12, 2021) "22 Celebrities that have tried Ayahuasca." *New Life Ayahuasca.*
11. Paturel, Amy. (January 2, 2019) "The Science Behind Coincidence." *Discover Magazine.*
12. Wynn, Stephen; Wynn, Tanya. (2017) *Women in the Great War*. Pen & Sword Books.
13. Falk Moore, Sally. (August 12, 1984) "Sacrificing the Boy." *The New York Times.*
14. (2017) "Edgar Allan Poe Invents the Modern Detective Story." *National Park Service.gov.*
15. Chen, Edwin. (December 9, 1979) "Twins Reared Apart: A Living Lab." *The New York Times.*
16. Kerr, Jolie. (November 17, 2018) "How to Talk to People." *The New York Times.*

17. Gormly, Kellie B. (March 29, 2022) "How Kate Warne, America's First Woman Detective, Foiled a Plot to Assassinate Abraham Lincoln." *Smithsonian Magazine.*
18. (2022) "Women Over 50: The Right To Be Seen on Screen." *See Jane.org.*

Conclusion

1. Abbey, Alison. (September 14, 2020) "15 Agatha Christie Quotes to Celebrate the Author's 130th Birthday." *Parade Magazine.*

Index